W9-APX-900

An American National Standard

IEEE Recommended Practice for Emergency and Standby Power Systems for Industrial and Commercial Applications

Published by
The Institute of Electrical and Electronics Engineers, Inc

Distributed in cooperation with
Wiley-Interscience, a division of John Wiley & Sons, Inc

ANSI/IEEE
Std 446-1987
(Revision of ANSI/IEEE
Std 446-1980)

An American National Standard

IEEE Recommended Practice for Emergency and Standby Power Systems for Industrial and Commercial Applications

Sponsor

Power Systems Engineering Committee
of the
Industrial and Commercial Power Systems Department
of the
IEEE Industry Applications Society

Approved June 19, 1986

IEEE Standards Board

Approved November 17, 1986

American National Standards Institute

First Printing
February 1987

ISBN 471-62571-X

Library of Congress Catalog Number 86-46292

© Copyright 1987 by

The Institute of Electrical and Electronics Engineers, Inc
345 East 47th Street, New York, NY 10017, USA

No part of this publication may be reproduced in any form,
in an electronic system or otherwise,
without the prior written permission of the publisher.

February 13, 1987

SH10835

Foreword

(This Foreword is not a part of ANSI/IEEE Std 446-1987, IEEE Recommended Practice for Emergency and Standby Power Systems for Industrial and Commercial Applications.)

In 1968 the Industrial and Commercial Power Systems Committee within the Industry and General Applications Group of the Institute of Electrical and Electronics Engineers recognized that a need existed for a publication that would provide guidance to industrial users and suppliers of emergency and standby power systems.

The nature of electric power failures, interruptions, and their duration covers a range in time from microseconds to days. Voltage excursions occur within a range from 20 times normal (or more) to a complete absence of voltage. Frequency excursions can vary as widely in many forms, from harmonics to direct current. These variables occur due to a multitude of conditions both in the power system ahead of the point of the user's service entrance and following the service entrance within the user's area of distribution.

Such elements as lightning, automobiles striking power poles, ice storms, tornadoes, switching to alternate lines, and equipment failure are but a few of the causes of interruptions in the electric power supply ahead of the service entrance.

Within the user's area of distribution are such elements as short and open circuits, undersized feeders, equipment failures, operator errors, temporary overloads, single-phasing unbalanced feeders, fire, switching, and many other generators of power interruption or failure.

In the past the demand for reliable electric power was less critical. If power was completely interrupted too often, another source was found. If voltage varied enough to cause a problem, a regulator or a larger conductor was installed. As processes, controls, and instrumentation became more sophisticated and interlocked, the demand developed to shorten the length of outages. Increased safety standards for people required emergency and exit lighting. Many factories added medical facilities that needed reliable electric power.

With the advent of solid-state electronics and computers, the need for continuous, reliable, high-quality electric power became critical. Many installations required uninterruptible power, virtually free of frequency excursions and voltage dips, surges, and transients.

In 1969 a working group was established under the Industrial Plants Power Systems Subcommittee of the Industrial and Commercial Power Systems Committee to collect data and produce a publication entitled "Emergency Power Systems for Industrial Plants." Later that year the scope of the work was enlarged to include standby power since in meeting various needs the two systems were often found to be intertwined, or one system served multiple purposes.

As the work progressed it became apparent that industrial and commercial needs contained more similarities than differences. Systems available to supply the required power to industry were found applicable to both fields. Once again the scope of the work was expanded to include commercial requirements. The working group was changed to the status of a subcommittee under the Industrial and Commercial Power Systems Committee, and the proposed publication was directed toward establishing recommended practices. As a result of additional organizational changes, the Emergency and Standby Power Systems Subcommittee is now under the Power Systems Engineering Committee of the Industrial and Commercial Power Systems Department.

This second revision of the IEEE Orange Book includes new and updated material. The sections on maintenance, protection, grounding, and industry applications have been revised and expanded to cover a new and broader scope. In addition, the previous chapter on systems and hardware has been divided into two new chapters providing a more logical separation and presentation of equipment.

This IEEE Recommended Practice continues to serve as a companion publication to the following other Recommended Practices prepared by the IEEE Industrial and Commercial Power Systems Committee:

Recommended Practice for Electric Power Distribution for Industrial Plants (IEEE Red Book), ANSI/IEEE Std 141-1986.

Recommended Practice for Grounding of Industrial and Commercial Power Systems (IEEE Green Book), ANSI/IEEE Std 142-1982.

Recommended Practice for Electric Power Systems in Commercial Buildings (IEEE Gray Book), ANSI/IEEE Std 241-1983.

Recommended Practice for Protection and Coordination of Industrial and Commercial Power Systems (IEEE Buff Book), ANSI/IEEE Std 242-1986.

Recommended Practice for Industrial and Commercial Power System Analysis (IEEE Brown Book), ANSI/IEEE Std 399-1980.

Recommended Practice for the Design of Reliable Industrial and Commercial Power Systems (IEEE Gold Book), ANSI/IEEE Std 493-1980.

Recommended Practice for Electric Systems in Health Care Facilities (IEEE White Book), ANSI/IEEE Std 602-1986.

Recommended Practice for Energy Conservation and Cost Effective Planning in Industrial Facilities (IEEE Bronze Book), ANSI/IEEE Std 739-1984.

At the time this recommended practice was approved, the Emergency and Stand-by Power Systems Subcommittee had the following members and contributors:

Pat O'Donnell, *Chairman*

Graydon M. Bauer	Thomas S. Key	Farrokh Shokooh
Richard Bowyer	Don Koval	Myron A. Smith
Rene Castenschiold	An T. Le	L. H. Soderholm
Kao Chen	A. C. Lordi	J. Charles Solt
Patrick L. Daigle	Charles R. McDonald	Jay Stewart
Richard M. Donnell	Neil Nichols	Georg Stromme
William R. Haack	Charles D. Potts	Max D. Trundle
G. L. Hensel	Jack Ripley	Eli Yagor
Gordon S. Johnson	Kurt Shafer	Donald W. Zipse

The following persons were on the balloting committee that approved this document for submission to the IEEE Standards Board:

Graydon M. Bauer	Don Koval	Farrokh Shokooh
C. E. Becker	A. C. Lordi	Myron A. Smith
Rene Castenschiold	B. K. Mathur	L. H. Solderholm
Kao Chen	Charles R. McDonald	J. Charles Solt
Patrick L. Daigle	R. H. McFadden	Jay Stewart
William R. Haack	W. J. Neiswender	Georg Stromme
C. R. Heising	Neil Nichols	Max D. Trundle
G. L. Hensel	Pat O'Donnell	Eli Yagor
Gordon S. Johnson	Charles D. Potts	Donald W. Zipse
Thomas S. Key	E. Rappaport	M. Zucker
	Jack Ripley	

When the IEEE Standards Board approved this standard on June 19, 1986, it had the following membership:

John E. May, *Chairman*   **Irving Kolodny,** *Vice Chairman*
Sava I. Sherr, *Secretary*

James H. Beall	Jack Kinn	Robert E. Rountree
Fletcher J. Buckley	Joseph L. Koepfinger*	Martha Sloan
Paul G. Cummings	Edward Lohse	Oley Wanaselja
Donald C. Fleckenstein	Lawrence V. McCall	J. Richard Weger
Jay Forster	Donald T. Michael*	William B. Wilkens
Daniel L. Goldberg	Marco W. Migliaro	Helen M. Wood
Kenneth D. Hendrix	Stanley Owens	Charles J. Wylie
Irvin N. Howell	John P. Riganati	Donald W. Zipse
	Frank L. Rose	

*Member emeritus

IEEE Standards documents are developed within the Technical Committees of the IEEE Societies and the Standards Coordinating Committees of the IEEE Standards Board. Members of the committees serve voluntarily and without compensation. They are not necessarily members of the Institute. The standards developed within IEEE represent a consensus of the broad expertise on the subject within the Institute as well as those activities outside of IEEE which have expressed an interest in participating in the development of the standard.

Use of an IEEE Standard is wholly voluntary. The existence of an IEEE Standard does not imply that there are no other ways to produce, test, measure, purchase, market, or provide other goods and services related to the scope of the IEEE Standard. Furthermore, the viewpoint expressed at the time a standard is approved and issued is subject to change brought about through developments in the state of the art and comments received from users of the standard. Every IEEE Standard is subjected to review at least once every five years for revision or reaffirmation. When a document is more than five years old, and has not been reaffirmed, it is reasonable to conclude that its contents, although still of some value, do not wholly reflect the present state of the art. Users are cautioned to check to determine that they have the latest edition of any IEEE Standard.

Comments for revision of IEEE Standards are welcome from any interested party, regardless of membership affiliation with IEEE. Suggestions for changes in documents should be in the form of a proposed change of text, together with appropriate supporting comments.

Interpretations: Occasionally questions may arise regarding the meaning of portions of standards as they relate to specific applications. When the need for interpretations is brought to the attention of IEEE, the Institute will initiate action to prepare appropriate responses. Since IEEE Standards represent a consensus of all concerned interests, it is important to ensure that any interpretation has also received the concurrence of a balance of interests. For this reason IEEE and the members of its technical committees are not able to provide an instant response to interpretation requests except in those cases where the matter has previously received formal consideration.

Comments on standards and requests for interpretations should be addressed to:

Secretary, IEEE Standards Board
345 East 47th Street
New York, NY 10017
USA

Emergency and Standby Power Systems for Industrial and Commercial Applications

Working Group Members and Contributors

Pat O'Donnell, *Chairman*

Chapter 1 — Scope: Pat O'Donnell, *Chairman*; Rene Castenschiold, Georg Stromme, Donald W. Zipse

Chapter 2 — Definitions: Pat O'Donnell, *Chairman*; Rene Castenschiold, Georg Stromme, Donald W. Zipse

Chapter 3 — General Need Guidelines: Donald W. Zipse, *Chairman*; Kao Chen, Charles R. McDonald, Jack Ripley, J. Charles Solt, Georg Stromme, Eli Yagor

Chapter 4 — Generator and Electric Utility Systems: Gordon S. Johnson, *Chairman*; Rene Castenschiold, Richard M. Donnell, Don Koval, Farrokh Shokooh

Chapter 5 — Stored Energy Systems: William R. Haack, *Chairman*; Richard Bowyer, Thomas S. Key, An T. Le, Charles D. Potts, Jack Ripley, Kurt Shafer, Max D. Trundle

Chapter 6 — Protection: Pat O'Donnell, *Chairman*; Rene Castenschiold, Kao Chen, Neil Nichols, Farrokh Shokooh, Georg Stromme

Chapter 7 — Grounding: Rene Castenschiold, *Chairman*; Patrick L. Daigle, Gordon S. Johnson, Neil Nichols, Georg Stromme, Donald W. Zipse

Chapter 8 — Maintenance: Charles D. Potts, *Chairman*; Graydon M. Bauer, Gordon S. Johnson, Charles R. McDonald, J. Charles Solt, Donald W. Zipse

Chapter 9 — Industry Applications: Neil Nichols, *Chairman*; Patrick L. Daigle, Pat O'Donnell, Jay Stewart, Max D. Trundle, Eli Yagor

Contents

Chapter 1
Scope

This standard presents recommended engineering principles, practices, and guidelines for the selection and application of emergency and standby power systems. Industrial and commercial users' needs are outlined and discussed and the material is primarily presented from a user's viewpoint. However, handling of the effects of power system disturbances obviously requires close cooperation between users, electric utilities, and equipment manufacturers, and the part played by each in the solutions to users' needs is vital to the proper selection and application of equipment.

Those responsible for reliable power for industrial and commercial operations are provided with itemized operational needs that form the basis for emergency and standby power system design and application. Industrial and commercial needs are sufficiently described, with a distinction made between the two with reference to mandatory laws, regulations, codes, and standards.

Chapters such as "General Need Guidelines" provide electric utility companies and equipment manufacturers with information on user requirements that allow for closer coordination in meeting specific power supply needs and proper equipment and system design for an efficient, practical, and reliable installation. Equipment manufacturers have provided technical and economic information on available hardware and systems, allowing recommendations for the requirements of various types of installations.

Technical guidelines in specialized areas of design such as protection and grounding, as specifically required in emergency and standby power systems, are presented in separate chapters. These chapters assist the designer in the actual application of equipment. Additional information on applications by specific industry types is also included in a separate chapter. This provides users and suppliers of emergency and standby power systems with vital information on how various industries meet their specific users' needs.

To round out the proper selection and application of emergency and standby power systems, proper maintenance guidelines must be included. Guidelines for specific types of equipment are discussed, recognizing that experience and economic factors play a large part in how various users must maintain equipment.

In this standard, the reader will find answers to the following questions about guidelines.

(1) Is an emergency or standby power system, or both, needed, and what will it accomplish?

(2) What types of systems are available and which can best meet the users' needs?

(3) What are the maintenance and operating requirements for maintaining system reliability?

(4) Where can additional information be obtained?

(5) How should the most suitable system be designed and applied to the existing power system?

The following industries and fields are considered specialized and beyond the scope of inclusion and direct address by this standard: transport, government, military, and utility.

Chapter 2
Definitions

2.1 Introduction. This chapter is intended to conveniently provide terms and definitions applicable to this standard for the purpose of aiding in its overall understanding. These terms and definitions are in no way intended to supersede, contradict, or change those provided in ANSI/IEEE Std 100-1984 [1].[1]

automatic controller (process control). A device that operates automatically to regulate a controlled variable in response to a command and a feedback signal.

automatic transfer switch. Self-acting equipment for transferring one or more load conductor connections from one power source to another.

availability. The fraction of time within which a system is actually capable of performing its mission.

bypass/isolation switch. A manually operated device used in conjunction with an automatic transfer switch to provide a means of directly connecting load conductors to a power source and of disconnecting the automatic transfer switch.

commercial power. Power furnished by an electric power utility company; when available, it is usually the prime power source. However, when economically feasible, it sometimes serves as an alternative or standby source.

computer. (1) A machine for carrying out calculations. (2) By extension, a machine for carrying out specified transformations on information.

current withstand rating. The maximum allowable current, either instantaneous or for a specified period of time, that a device can withstand without damage.

data processing. Pertaining to any operation or combination of operations on data.

data processor. Any device capable of being used to perform operations on data, for example, a desk calculator, tape recorder, analog computer, or digital computer.

---

[1] The numbers in brackets correspond to those of the references listed at the end of this chapter; when preceded by B, they correspond to the bibliography at the end of this chapter.

dropout voltage (or current). The voltage (or current) at which a magnetically operated device will release to its de-energized position. It is a level of voltage (or current) that is insufficient to maintain the device in an energized state.

emergency power system. An independent reserve source of electric energy that, upon failure or outage of the normal source, automatically provides reliable electric power within a specified time to critical devices and equipment whose failure to operate satisfactorily would jeopardize the health and safety of personnel or result in damage to property.

firm power. Power intended to be always available, even under emergency conditions.

forced outage. A power outage that results from the failure of a system component, requiring that it be taken out of service immediately, either automatically or by manual switching operations, or an outage caused by improper operation of equipment or human error. This type of power outage is not directly controllable and is usually unexpected.

frequency droop. The absolute change in frequency between steady-state no load and steady-state full load.

frequency regulation. The percentage change in emergency or standby power frequency from steady-state no load to steady-state full load:

$$\%R = \frac{F_{n1} - F_{f1}}{F_{n1}} \cdot 100$$

harmonic content. A measure of the presence of harmonics in a voltage or current waveform expressed as a percentage of the amplitude of the fundamental frequency at each harmonic frequency. The total harmonic content is expressed as the square root of the sum of the squares of each of the harmonic amplitudes (expressed as a percentage of the fundamental).

load shedding. The process of deliberately removing preselected loads from a power system in response to an abnormal condition in order to maintain the integrity of the system.

off-line operation. (1) Pertaining to computer systems not under direct control of the central processing unit. (2) Pertaining to uninterruptible power supply systems whereby an inverter is off during normal operating conditions.

on-line operation. (1) Pertaining to equipment or devices under direct control of the central processing unit. (2) Pertaining to uninterruptible power supply systems whereby an inverter is on during normal operation conditions.

power failure. Any variation in electric power supply that causes unacceptable performance of the user's equipment.

power outage. Complete absence of power at the point of use.

prime mover. The machine used to develop mechanical horsepower to drive an emergency or standby generator to produce electrical power.

prime power. That source of supply of electrical energy that is normally available and used continuously day and night, usually supplied by an electric utility company, but sometimes supplied by base-loaded user-owned generation.

real time (processing). Pertaining to the actual time during which a physical process transpires or pertaining to the performance of a computation during the actual time of related physical processing in order that results of the computation can be used in guiding the physical process.

redundancy. Duplication of elements in a system or installation for the purpose of enhancing the reliability or continuity of operation of the system or installation.

scheduled outage. A power outage that results when a component is deliberately taken out of service at a selected time, usually for purposes of construction, preventive maintenance, or repair. This type of outage is directly controllable and usually predictable.

separate excitation. A source of generator field excitation power derived from a source independent of the generator output power.

shunt excitation. A source of generator field excitation power taken from the generator output, normally through power potential transformers connected directly or indirectly to the generator output terminals.

standby power system. An independent reserve source of electric energy that, upon failure or outage of the normal source, provides electric power of acceptable quality so that the user's facilities may continue in satisfactory operation.

transient. That part of the change in a variable, such as voltage or amperage, that disappears during transition from one steady-state operating condition to another.

uninterruptible power supply (UPS). A system designed to automatically provide power, without delay or transients, during any period when the normal power supply is incapable of performing acceptably.

utility power. Synonym: **commercial power.**

2.2 References. The following publication shall be used in conjunction with this chapter.

[1] ANSI/IEEE Std 100-1984, IEEE Standard Dictionary of Electrical and Electronics Terms.[2]

---

[2] ANSI/IEEE publications can be obtained from the Sales Department, American National Standards Institute, 1430 Broadway, New York, NY 10018, or from Publication Sales, the Institute of Electrical and Electronics Engineers, Service Center, Piscataway, NJ 08854.

Chapter 3
General Need Guidelines

3.1 Introduction. While everyone who uses electric power desires perfect frequency, voltage stability, and reliability at all times, this cannot be realized in practice. Within a complex facility, the requirements are continually changing and becoming more demanding and interlocked.

The supplying utility cannot be expected to provide a perfect power supply because many of the causes of power supply disturbances are beyond the control of the utility. For example, automobiles hit poles, animals climb across insulators, lightning strikes overhead lines, and cyclones, hurricanes, and high winds blow tree branches and other debris into lines.

Lightning, wind, and rain produced by thunderstorms cause power failures in the form of power interruptions and transients. Figure 1 is useful in determining the probability of such power failures depending upon the user's geographic location. However, most utilities or commercial power companies, through the use of proper shielding and surge arrestors, can provide service that is of equal reliability in any area.

Figure 2 shows the density of tornadoes in several states. These violent local storms take their toll on power systems, as do more widely located storms, such as hurricanes, ice and snow storms, and floods. The weighted consideration must be based on their frequency and severity in the user's location.

Although there is less chance for an interruption on an underground system, and although an underground system has fewer dips, the duration of any interruption that occurs may be much longer because of the longer time required to find out where a cable failure occurred and to make repairs.

Even the operation of protective devices can cause power supply disturbances. As an example, overcurrent and short-circuit protective devices require excessive current to operate and will be accompanied by a voltage dip on any line supplying the excessive current until the device opens to clear the fault. Devices opening to clear a lightning flashover with subsequent reclosing cause a momentary interruption.

Utility companies have little practical control over, and thus cannot accept the responsibility for, disturbances on their systems. This is illustrated by the text taken from a typical utility sales contract: "The power delivered hereunder shall be three-phase alternating current at a frequency of approximately 60 Hz and at a nominal

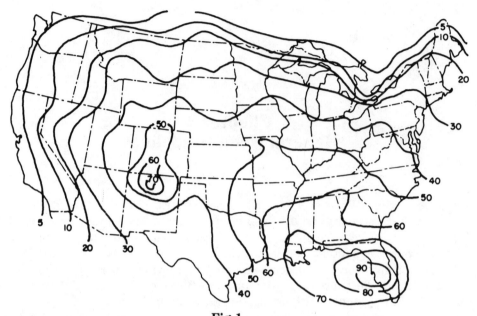

Fig 1
Average Number of Thunderstorm Days per Year

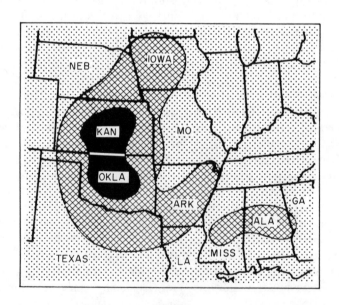

Fig 2
Approximate Density of Tornadoes
Cross-Hatched Areas — 100-200 in 40 Years
Black Area — Over 200 in 40 Years
(See [23], 3.15.1)

voltage of 208Y/120 V. Except for temporary periods of abnormal operating conditions, variations from normal voltage shall not exceed 5% up or down. The utility will use reasonable diligence to provide a regular and uninterrupted supply of power, but in case such supply should be interrupted for any cause, the utility shall not be liable for damages resulting therefrom."

It is obvious that utility companies are unwilling to make power quality or continuity guarantees, and rightly so.

In several instances, utility companies have recognized the limitations of their power quality and have offered auxiliary equipment with special purchase agreements to satisfy the needs of those who use sensitive utilization equipment.

Utility companies are by no means the only source of power system disturbances. Disturbances and outages also occur in the plant system through the loss of power due to short circuits in wiring and failures of local generation including emergency and standby equipment. *Noise* is generated in otherwise acceptable electric power by motors, welders, switches, semiconductor-controlled rectifier (SCR) gating, dielectric heating, arcing short circuits, and a myriad of other sources.

Ordinary power meters and instrumentation cannot be used to measure power transients or disturbances. Recording devices with extremely fast response must be used to detect, measure, and record disturbance magnitudes and durations. Most disturbances will not only vary within a given 24 h period, but depending on the geographical area, could be subject to seasonal variations. Usually, there are no simple methods available to record, diagnose, and classify the nature of the power problems.

Any meaningful data with regard to power quality, or lack of it, will be obvious only after conducting a thorough and detailed measurement recording and analysis program. Such a program would include the monitoring of an incoming power line over a representative period of time. Since much of the instrumentation is costly (typically $3000–$16 000) in relation to its limited duration usage, rental arrangements are often a popular choice.

Since ideal power quality and continuity can seldom be obtained from the supplying utility, this text has been prepared to indicate how the effects of power supply disturbances may be reduced to acceptable levels, or even eliminated. These reduction methods include

(1) Modification of the design of utilization equipment so as to be impervious to power disturbances and discontinuities

(2) Modification of the prime power distribution system to be compatible with utilization equipment

(3) Modification of both systems and equipment to meet a criterion that is realistic for both

(4) Interposing a continuous electric supply system between the prime source and the utilization equipment; this will function as a *buffer* to external sources of transients, but depending on the design, could increase the magnitude of load induced disturbances

Almost any significant deviation from normal power parameters may be capable of causing problems with some electronic equipment. Steady or slow deviations that exceed the product design range of line voltage or frequency can affect the shaft speed of motors, the force and speed of actuators, and the conversion of alternating voltage into regulated dc voltage for electronic circuit operation. As line voltage approaches product design limits, this often reduces the ability of the product to sustain, without incident, a line voltage transient as described in the following. Of course, some electronic circuits are more susceptible than others. This depends upon specific application and design.

Most frequent among excursions from normal line conditions are those classed as voltage transients. These often contain an initial fast voltage rise or fall time (sometimes oscillatory) followed by a slower, longer duration change. Thus one transient event may contain both fast impulses or noise and a slower change in voltage.

Fast voltage impulses and noise generally have little direct effect upon the smooth, even flow of power to such devices as motors and ac to dc power conversion equipment. The flywheel effect of motors and the energy storage in filters for reducing radio frequency interference and ripple in rectified dc output current of ac to dc converters keep the fast voltage changes from affecting their outputs directly. Excessively large impulses or noise bursts and inadequate filtering may result in component voltage overstress or in the premature triggering of control circuit elements such as SCRs and TRIACs. Otherwise, very short duration impulses and high-frequency noise have little effect upon electronic circuits via power supply paths. More often these disturbances reach sensitive electronic circuits by more subtle paths through circuit grounds where power and signal ground circuit paths intersect. This dictates that care be taken in the location of connections and routing of ground conductors in addition to the more obvious precautions taken to reduce the number and amplitude of transients.

Without a good power disturbance monitor to measure and record disturbances continuously, there is a tendency to make judgments regarding disturbances in the power source in terms of their effects upon various electrical loads. When the lights go out and the electrical equipment all stops working, this is usually a fairly good indication of a power interruption. However, if the lights merely flicker or the electronic equipment malfunctions, it is difficult to judge whether or not there has been a severe change in voltage for a very short time or a very small voltage change for a much longer time. One cannot determine without a disturbance monitor whether or not the disturbance was unusually severe or the electronic equipment was unusually susceptible to the disturbance. Without the necessary features in a disturbance monitor, one cannot tell if the source of the disturbance was external to the load equipment or was the load equipment itself.

For practical reasons and convenience, power disturbances should be defined in terms that are related to practical methods of measurement. When power disturbance limits are given in the specifications for electrical and electronic equipment, use of the same terminology and definitions used in measurement will make it easier to resolve problems and in some cases will help identify the responsibility.

For purposes of later discussions, disturbances in ac power may be classified as deviations in one or more of the following areas:

(1) Voltage
(2) Wave shape

(3) Frequency
(4) Phase relationships
Of these, the most frequently encountered deviations occur in voltage as follows:
(1) Steady-state values (slow average), including unbalance
(2) Outages and interruptions
(3) Surges and sags
(4) Impulses and noise

NOTE: The term *transients* applies loosely to items (3) and (4).

The steady-state value is measured with a true rms-actuated, rms-reading volt-meter or the equivalent. The instrument should be damped, or otherwise, the readings must be averaged over a 5–10 s averaging time so that a succession of readings will indicate the gradual, long-term changes in averaged rms voltage level. This is also known as slow-average rms voltage. Most deviations in steady-state voltage are caused by voltage drops in power lines, transformers, and feeders as load is increased. When step voltage corrections are made by such means as transformer tap changing or by adding or removing capacitors, transients will be generated at the time of the step change. In addition, the change that corrected an undervoltage condition as the load increased will have to be reversed later when the load decreases. Otherwise, excessive overvoltage may result. Voltage is usually changed gradually in many small incremental steps. Occasional large step changes can occur.

The worst steady-state voltage deviations are likely to occur in areas where the total load approaches and even exceeds the capacity to generate or distribute power. In spite of short-time overload ratings, it becomes necessary to reduce load during periods of peak loading such as during heat waves when air conditioning is added to all the other normal loads. Under such conditions, not only may there be an unusually large voltage drop between power source and load, but there may also be a planned voltage reduction to relieve utility system loading. This voltage reduction is known as a *brownout*. With reductions of 3%, 5%, or even 8% at points where voltage is regular, the voltage reduction at load points may be an additional 5–10%.

The most common protection against brownouts is some form of voltage regulation, preferably one that has sufficiently low internal impedance and fast response time to avoid disturbances created by load changes and by phase controlled load regulation (where used).

When a planned brownout fails to relieve the utility power system overload enough to supply energy at reduced voltage to all users, one remaining alternative is to shed loads in a rotating sequence known as a *rolling blackout*. Under some emergency plans, certain noncritical loads would be shed for the duration of the overload. A few known critical loads would not be disturbed. The remainder would be subjected to power interruptions lasting 10–20 min each on a rotating basis among groups of subscribers, sufficient to relieve the overload by means of the rolling blackout.

Unequal loading on polyphase lines or single-phase three-wire lines is often the cause of voltage imbalance. Voltage control devices generally operate to regulate the average of the phase voltages, or sometimes just one of the phase voltages, on the assumption that the other phases will be equal to it. The most common corrective measure is to balance the loading among the phases. However, power sources with high internal impedance are more critical in a requirement to distribute loads evenly in order to avoid excessive voltage imbalance. Selection of a delta-connected primary with a wye-connected secondary rather than a four-wire wye input and output will help distribute unequal phase loading.

Large computer installations with an uninterruptible power supply (UPS) capable of supplying the computer's power needs for 5–15 min would need supplementary standby power to operate auxiliary building services, such as air conditioning, and to replenish the storage battery energy in order that the computer system could continue to remain functional through an extended power outage. Diesel- or gas-turbine-driven generators fill this need.

Nonpermanent departures from the normal line voltages and frequency can be classified as disturbances. Disturbances include impulses, noise, transients, and even some changes in frequency or sudden phase shifts during synchronizing operations. Although the frequency and phase shift events may rarely be encountered when large loads are switched in power networks, these may frequently be encountered in small independent power sources during synchronizing and switching of loads from one source to another.

The more usual disturbance involves line voltage impulses, noise, transients, steady-state voltage change, or some combination of these. Most disturbances on a power system are of short duration. Studies indicate that 90% are less than 1 s in duration. Some 80–85% involve only one phase of a three-phase system. The following is an attempt to group disturbances in relation to their duration and possible causes.

3.1.1 1 s–1 min. Usually attributed to severe faults accompanies by 50–100% voltage loss on one or more phases, these disturbances often result in an outage on some circuit. Faults often involve all three phases and may be the result of a downed pole, a tree or crane on the line, a breaker lockout, or an in-line fuse blowing. If the critical load is on the cleared side of the fault, the disturbance becomes an outage. If it is on the power source side of the fault clearing device, the normal voltage may be restored.

3.1.2 10–40 Cycles. The disturbances are surges and sags due to operation of relatively slow-speed breakers, reclosures in clearing faults on adjacent circuits, tap changing of in-line transformers and regulators, and the starting of motors, either across the line or reduced voltage start.

3.1.3 0–8 Cycles. The disturbances are surges, but more often sags, caused by a fault and subsequent fuse action or high-speed circuit breaker operation on adjacent circuits. Switching surges and inrush current transients caused when energizing electrical devices create short-duration voltage sags and surges. Single-phase loads on a multiline power source (single- or three-phase) create voltage surges on the unloaded phases if they cause sags on the loaded phase.

3.1.4 0.001–1 Cycles. These disturbances are short-duration surges and sags caused by lightning arresters, load and capacitor switching, and short-duration faults. Any disturbances lasting less than 1 cycle may be difficult to compare directly with those lasting longer. Measurement of voltages for periods longer than one cycle is conveniently done in terms of rms values (the root mean square of all values during the cycle). Since most utilization voltages are supplied via a transformer in series with the power source, it is unlikely that a voltage surge at its output will exceed 130–150% of nominal volt-seconds (the area under a half sine wave). Core saturation in the transformer tends to limit this. Greater input volt-seconds that would push the magnetic core into deeper saturation would result in overcurrent tripping. The net result is that the transformer can pass very high voltages if they are of short duration, but the longer they persist, the lower the voltage must be.

3.1.5 Less than 0.001 Cycle. These disturbances are generally classed as impulses. (They originate as the fast portions at the beginning or end of switching transients that may have much longer durations. They may be associated with disturbances of all kinds, the more severe of which are natural lightning, electrostatic discharge, and the switching of nearby loads sharing the same power feeder.)

Impulses may be of either polarity, single fast-rise and fast-fall time event, or they can be damped-oscillatory in form. There can be bursts of multiple impulses. They can be synchronized with the power frequency so as to occur at the peak of each voltage wave (arcing fault, for example), or they can occur at random.

Impulses are measured in most instrumentation systems as voltage peak deviation from the sine wave voltage. This is usually accomplished by use of a high-pass filter (simple capacitor–resistor combination) to block the power frequency voltage. If a positive-going impulse occurs at the positive peak of a sine wave, the maximum instantaneous voltage will be the sum of the two. If not measured or expressed in volts, impulse amplitude limits are expressed in terms of the nominal voltage peak. For example, an impulse of *3 times nominal* referred to a 120 V rms nominal voltage would imply an impulse 3 times as great as the peak value of the sine wave. This is $\sqrt{2} \cdot 120 \cdot 3 = 509$ V peak value for the impulse. If this occurred at the peak of the sine wave and was in the same direction, the two would add to give $509 + (\sqrt{2} \cdot 120) = 679$ V peak.

A fast impulse applied to one conductor will tend to couple to the other nearby conductors as it travels. Its amplitude and rise time will also decrease as it travels. By the time that an impulse travels more than 20 or 30 ft, it is likely that there will be considerable common mode signal in the conductor bundle unless the signal started as a balanced signal over a balanced line (equal positive and negative pulse signals on each of a balanced wire pair). In many sensitive electronic units, common mode impulses and high-frequency noise from power source disturbances are a major cause of malfunctions. Isolating transformers placed close to the electronic equipment and the physical routing of grounding conductors to minimize intersection of power and low-voltage level signal ground paths will often reduce or eliminate power-related transient noise problems.

Electrical noise can have much the same effect as impulses upon sensitive electronic equipment and requires similar measures to cope with it. In digital equipment, digital circuits generally operate in one of two (binary) states. Its circuits are relatively insensitive to external disturbances most of the time. However, at transition and clock times, a noise signal that otherwise could not influence a logic state might alter it and corrupt the signal. The statistical probability may be small but still create sufficient trouble to make a computer operation unacceptable to a user. Impulses are generally discrete events, and if there are not too many of them in a day, they may escape notice as a cause of difficulty. Noise, on the other hand, may be more continuous or occur in bursts. Depending upon occurrence and intensity, noise may be a more severe disturbance in terms of the number and frequency of malfunctions it causes.

Noise may be random in frequency content, but is more likely to have a strong frequency structure, some components of which may be at high radio frequencies. If power and grounding conductor lengths coincide with wavelengths (or multiples of quarter wavelengths) of noise frequencies, resonant conditions may occur that will greatly amplify the noise signals.

Grounding is an essential part of power sources and their connections to loads. In addition to making ground conductors comply with local codes for safety to personnel (keeping exposed metal at ground potential) and providing low impedance return paths so that circuit protection will clear faults, grounding should also serve as a constant potential signal reference from one part of the grounding system to another over a broad frequency range.

For much electrical equipment, 70–90% voltage protection is adequate, that is, transfer to emergency power when the line voltage drops to 70% of normal and retransfer when it returns to 90%. When specialized loads are used, the manufacturer's recommendations should be followed. For motor loads, the protection can be increased to provide 80–90% protection. Too close a voltage spread will cause many false or unnecessary transfers; too wide a spread will cause equipment damage and malfunctions. The length of time of the voltage dip is of paramount importance.

Justification of the expenditures necessary to fulfill *user needs* for emergency and standby power systems falls into three broad classifications:

(1) Mandatory installations to meet laws and regulations of federal, state, county, and city

(2) Maintenance of the safety and health of people involved during a power failure

(3) Increased profits due to fewer and shorter power failures

Power for the safety of personnel and to prevent pollution of the environment may become increasingly mandatory. A continual check of laws and regulations should be maintained to be current with the requirements. Public Law Number 91-596, Williams–Steiger Occupational Safety and Health Act of 1970 (OSHA) [25],[3] may materially alter the electric power reliability and availability requirements as the full interpretation and application of its provisions are published.

Table 1 is a guide to state codes and regulations for emergency power systems in the United States. All the latest codes and regulations for the area in which the industrial or commercial facility is located must be consulted and followed.

---

[3] The numbers in brackets correspond to those of the references listed at the end of this chapter; when preceded by B, they correspond to the bibliography at the end of this chapter.

Table 1
Codes for Emergency Power by States and Major Cities (Completed September 1984)

State/City	Does State/City Have Legislation?	Legislation Code Type?	Hospitals	Nursing Homes	Schools	Theaters (Public Gathering Places)	Office Buildings	Hotels	Apartment Buildings	Airports	Fire and Police Stations	Water Treatment Plants	Sewage Treatment Plants	All Public Buildings, State	All Public Buildings, Commercial	Applicable Government Agency	Smokeproof Enclosures in High-Rise Buildings
Alabama	Yes	4,6	A,C,D	A,C,D	C,D	C,D	C,D	C,D	C,D	C,D	C	C	C	C,D	C,D	M,N,O	B
Birmingham **	Yes	4,6	A,C,D	A,C,D	C,D	C,D	C,D	C,D	C,D	C,D	C			C,D	C,D	Q,S	A,B
Mobile **	Yes	1,4	A,C,D	A,C,D	C,D	A,C,D	A,C,D	A,C,D	A,C,D	C,D	C			C,D	C	S	
Alaska	Yes	1,3	A,C,D	A,C,D	A,C,D	A,C,D	A,C,D	A,C,D	A,C,D	A,C,D	C,D	C,D	C,D	C,D	A,C,D	M,P	B
Arizona	No																
Phoenix **	Yes	1	A,C,D	A,C,D	C,D	C,D	C,D	C,D	A,C,D	A,C,D	C,D	C,D	C,D	C,D	C,D	O	
Arkansas	Yes	1,6	A,C,D	A,C,D	C,D	A,C,D	A,C,D	A,C,D	A,C,D	C,D				C,D	A,C,D	M	B
California	Yes	2,3,4	A,C,D	A,C,D	C,D	C,D	C,D	C,D	C,D		C,D		C,D	C,D	C,D	O,T	A
Anaheim	Yes	2,3,4,7	A,C,D	A,C,D												M,N,O,Q	
Berkeley **	Yes	1,3	A,C,D	A,C,D	C,D	C,D	C,D	C,D	C,D	C,D	C,D	A,C,D	C,D	C,D	C,D	M,Q	B
Fresno	Yes	3,4	A,C,D	A,C,D	A,C,D	A,C,D	A,C,D	A,C,D	A,C,D	C,D	C,D	C,D	C,D	C,D	C,D	O,S	B
Glendale	Yes	3,4	A,C,D	A,C,D	A,C,D	A,C,D	A,C,D	A,C,D	A,C,D	C,D	C,D	C,D	C,D	C,D	C,D	M,Q	
Long Beach **	Yes	3	A,C,D	A,C,D	A,C,D	A,C,D	C,D	C,D	C,D	A,C,D	A,C,D	C,D	C,D	C,D	C,D	M,O	B
Los Angeles	Yes	3,4,8	A,C,D	A,C,D	A,C,D	A,C,D	A,C,D	A,C,D	A,C,D	B,C,D	C,D	C,D	C,D	C,D	C,D	M,Q,S	B
Oakland	Yes	1,3,4,8	A,C,D	A,C,D	A,C,D	A,C,D	A,C,D	A,C,D	A,C,D	C,D	C,D	C,D	C,D	C,D	C,D	M,O	B
Pasadena	Yes	1,2,3,4	A,C,D	C,D	C,D	A,C,D	C,D	C,D	C,D	C,D	C,D	C,D	C,D			O,Q	
San Diego	Yes	3,4	A,C,D	A,C,D	A,C,D	A,C,D	A,C,D	A,C,D	A,C,D	C,D	C,D	C,D	C,D	C,D	C,D	S	B
San Francisco	Yes	3,4	A,C,D	A,C,D	C,D	A,C,D	A,C,D	A,C,D	A,C,D	A,C,D	C,D	C,D	C,D	C,D	C,D	O	B
San Jose **	Yes	4														M,S	
Santa Ana **	Yes	3,4														M,Q	
Colorado	Yes	3,4	A,C,D	A,C,D	C,D	C,D	A,C,D	A,C,D	A,C,D	C,D			C,D	C,D	C,D	Q	B
Denver	Yes	4,8	A,C,D	A,C,D	C,D	A,C,D	A,C,D	A,C,D	A,C,D	A,C,D			C,D	A,C,D	A,C,D	Q	B
Connecticut	Yes	2,4	A,C,D	A,C,D	A,C,D	A,C,D	C,D	C,D	C,D	C,D	C,D	C,D	C,D	C,D	C,D	P	A
Hartford **	Yes	2	A,C,D	A,C,D	A,C,D	A,C,D	A,C,D	A,C,D	A,C,D	A,C,D	A,C,D	A,C,D	A,C,D	A,C,D	A,C,D	M,R	
New Haven **	Yes	1,2,4	A,C,D	A,C,D	A,C,D	A,C,D	A,C,D	A,C,D	A,C,D	C,D	A,C,D	A,C,D	A,C,D	A,C,D	A,C,D	M,Q	
Delaware	Yes	1,2,4	A,C,D	A,C,D	C,D	C,D	C,D	C,D	C,D	C,D				C,D	C,D	M	A

NOTE: An explanation of the numbers and letters used is given at the end of the table (see p 42).
Table 1 courtesy of the Electrical Generating Systems Marketing Association (April 1975).

Table 1 (Continued)

State/City	Does State/City Have Legislation?	Legislation Code Type?	Hospitals	Nursing Homes	Schools	Theaters (Public Gathering Places)	Office Buildings	Hotels	Apartment Buildings	Airports	Fire and Police Stations	Water Treatment Plants	Sewage Treatment Plants	All Public Buildings, State	All Public Buildings, Commercial	Applicable Government Agency	Smokeproof Enclosures in High-Rise Buildings
District of Columbia	Yes	2,4	A,C,D	A,C,D	C,D	C,D	C,D	C,D	C,D	C,D	C,D	C,D	C,D	C,D	C,D	M,Q	A
Florida	Yes	1,2,4,6	A,C,D	A,C,D		C,D	C,D	C,D	C,D	C,D	C,D	C,D	C,D	C,D	C,D	M,O	A
Jacksonville	Yes	8	A,C,D	A,C,D	C,D	C,D	C,D	C,D	C,D	C,D	C,D	C,D	C,D	C,D	C,D	Q,S	
St Petersburg **	Yes	1,4,7	A,C,D	A,C,D	A,C,D	A,C,D	A,C,D	A,C,D	A,C,D	A,C,D	A,C,D			A,C,D	A,C,D	U	A
Tampa	Yes	1,4,8	A,C,D	A,C,D	C,D	C,D	C,D	C,D	C,D	C,D	C,D			C,D	C,D	M	
Georgia	Yes	1,4,7	A,C,D	A,C,D	C,D	C,D	C,D	C,D	C,D	C,D	C,D			C,D	C,D	M	
Atlanta	Yes	1,4,8	A,C,D	A,C,D	C,D	C,D	C,D	C,D	C,D	C,D	C,D			C,D	C,D	O	B
Columbus **	Yes	4,6	A,C,D	B,C,D	B,C,D	B,C,D	C,D	B,C,D	C,D	C,D	C,D			C,D	C,D	S	
Savannah	Yes	4	A,C,D	A,C,D	C,D	C,D	C,D	C,D	C,D	C,D	C,D	C,D	C,D	C,D	C,D	S	
Hawaii	Yes	1,3	A,C,D	A,C,D	A,C,D	A,C,D	C,D	C,D	C,D	C,D	A,C,D	C,D	C,D	C,D	C,D	M,Q	B
Honolulu	Yes	1,3,8	A,C,D	A,C,D	C,D	C,D	C,D	C,D	C,D	C,D	C,D	C,D	C,D	C,D	C,D	M,Q	B
Idaho	Yes	1,3,4	A,C,D	A,C,D	C,D	C,D	C,D	C,D	C,D	C,D	C,D	B	B	C,D	C,D	M,R	B
Illinois	No	2	A,C,D	A,C,D	C,D	A,C,D	A,C,D	A,C,D	A,C,D	A,C,D	A,C,D	A,C,D	A,C,D	A,C,D	A,C,D	M,N	
Chicago **	Yes	8	A,C,D	A,C,D	A,C,D	A,C,D	A,C,D	A,C,D	A,C,D	A,C,D	A,C,D	A,C,D	A,C,D	A,C,D	A,C,D	S	B
Rockford **	Yes	1,4	A,C,D	A,C,D	A,C,D	A,C,D	A,C,D	A,C,D	A,C,D	A,C,D	A,C,D	C,D	C,D	A,C,D	A,C,D	S	B
Indiana	Yes	2,3,4	A,C,D	A,C,D	A,C,D	A,C,D	A,C,D	A,C,D	A,C,D	A,C,D	A,C,D	A,C,D	A,C,D	A,C,D	A,C,D	M,Q,R	B
Evansville	Yes	3,4	A,C,D	A,C,D	A,C,D	A,C,D	C,D	A,C,D	C,D	A,C,D	A,C,D			C,D	A,C,D	Q	B
Fort Wayne	Yes	1,3,4	A,C,D	A,C,D	A,C,D	A,C,D	A,C,D	A,C,D	C,D	A,C,D	A,C,D	B,C,D	B,C,D	A,C,D	A,C,D	M,Q	B
Gary	Yes	1,4	A,C,D	A,C,D	A,C,D	C,D	C,D	C,D	C,D	C,D	C,D	C,D	C,D	C,D	C,D	S	
Indianapolis **	Yes	2	A,C,D	A,C,D	A,C,D	A,C,D	A,C,D	C,D	C,D	A,C,D	A,C,D			A,C,D	A,C,D	S	B
South Bend	Yes	1,3,4	A,C,D	A,C,D	A,C,D	A,C,D	A,C,D	A,C,D	C,D	A,C,D	A,C,D	C,D	C,D	A,C,D	A,C,D	M	B
Iowa	Yes	1,4	A,C,D	A,C,D	C,D	C,D	C,D	C,D	C,D	C,D	C,D			C,D	C,D	M,Q	B
Des Moines	Yes	3,4	A,C,D	A,C,D	C,D	C,D	C,D	C,D	C,D	C,D	C,D			C,D	C,D	M,Q	B
Kansas	Yes	1,3,4	A,C,D	A,C,D	C,D	C,D	C,D	C,D	C,D	C,D	C,D					M,O	B
Kansas City **	Yes	3,4	A,C,D	A,C,D	C,D											S	B
Wichita **	Yes	4	A,C,D	A,C,D	C,D											S	B
Kentucky	Yes	1,2,4,5	A,C,D	A,C,D	C,D	C,D	C,D	C,D	C,D	C,D	C,D			C,D	C,D	O,Q	B

NOTE: An explanation of the numbers and letters used is given at the end of the table (see p 42).

Table 1 *(Continued)*

State/City	Does State/City Have Legislation?	Legislation Code Type?	Hospitals	Nursing Homes	Schools	Theaters (Public Gathering Places)	Office Buildings	Hotels	Apartment Buildings	Airports	Fire and Police Stations	Water Treatment Plants	Sewage Treatment Plants	All Public Buildings, State	All Public Buildings, Commercial	Applicable Government Agency	Smokeproof Enclosures in High-Rise Buildings
Louisiana	Yes	1,4	A,C,D	A,C,D	C,D	C,D	C,D	C,D	C,D	C,D	A,C,D	B	B	C,D	C,D	M,Q	
Baton Rouge **	Yes	4	A,C,D	A,C,D	B,C,D	A,C,D	B,C,D	A,C	C	A,C,D	A,C,D	B			C,D	S	A
New Orleans **	Yes	4	A,C,D	A,C,D	C,D	C,D	A,C,D*	A,C,D*	C,D	C,D	C,D		C,D	C,D	C,D	S	A
Shreveport	Yes	1,4	A,C,D	A,C,D	B,C,D	C,D	A,C,D	C,D	C,D	A,C,D	A,C,D	C,D	C,D	C,D	A,C,D	S	
Maine	Yes	1,2,4	A,C,D	C,D	C,D	C,D	C,D	C,D	C,D	C,D	C,D			C,D	C,D	M,S	
Maryland	Yes	1,2,4,5	A,C,D	A,C,D	C,D	C,D	C,D	C,D	C,D	A,C,D	C,D				C,D	M,N	
Baltimore **	Yes	4,8	A,C,D	C,D	C,D	A,C,D	C,D	A,C,D	C,D	A,C,D	A,C,D	C,D	C,D	C,D	C,D	S	
Massachusetts	Yes	2,5	A,C,D	A,C,D	A,C,D	C,D	C,D	C,D	C,D	C,D	A,C,D	C,D	C,D	C,D	C,D	P,Q	B
Bedford **	Yes	2	A,C,D	C,D	C,D	C,D	C,D	C,D	C,D	C,D	C,D	C,D	C,D	C,D	C,D	S	
Boston **	Yes	2	A,C,D	A,C,D	A,C,D	A,C,D	A,C,D	A,C,D	A,C,D	C,D				C,D		S	
Cambridge **	Yes	2	A,B	A,C,D	A,C,D	A,C,D	A,C,D	A,C,D	A,C,D	C,D	A,C,D	A,C,D	A,C,D	A,C,D	A,C,D	Q	
Springfield **	Yes	2	A,C,D	A,C,D	A,C,D	A,C,D	A,C,D	A,C,D	A,C,D		A,C,D			A,C,D	A,C,D	Q	
Worcester **	Yes	2	A,C	A,C	A,C	A,C		A,C	A,C	A,C	A,C					R	
Michigan	Yes	2,4,5	A,C,D	A,C,D	A,C,D	A,C,D	C,D	A,C,D	C,D	A,C,D	C,D			C,D	C,D	M,S	
Detroit	Yes	1,4,5	A,C,D	A,C,D	A,C,D	A,C,D	A,C,D	A,C,D	A,C,D	A,C,D	B,C,D	B,C,D	B,C,D		A,C,D	M,S	B
Flint **	Yes	4,5	A,C,D	A,C,D	A,C,D	A,C,D	A,C,D	A,C,D	A,C,D	A,C,D	A,C,D	B,C,D	B,C,D			M,S	B
Grand Rapids	Yes	1,4,5	A,C,D	A,C,D	A,C,D	A,C,D	C,D	A,C,D	C,D	C,D	C,D			C,D	D	S	B
Lansing **	Yes	2	A,C,D	A,C,D	A,C,D	A,C,D	A,C,D	A,C,D	A,C,D	A,C,D	A,C,D	B	B			S	
Minnesota	Yes	2,3,4,7	A,C,D	A,C,D	A,C,D	A,C,D	A,C,D	A,C,D	A,C,D	A,C,D	C,D			C,D	C,D	M,N,O	B
Minneapolis	Yes	1,3,4	A,C,D	A,C,D	A,C,D	A,C,D	A,C,D	A,C,D	A,C,D	A,C,D	C,D	C,D	C,D	C,D	C,D	M,S	B
Saint Paul	Yes	1,3,4	A,C,D	A,C,D	A,C,D	A,C,D	A,C,D	A,C,D	A,C,D	C,D	A	A,C,D	A,C,D	A,C,D	A,C,D	M,S	B
Mississippi	Yes	1,4,6	A,C,D	C,D	C,D	C,D	A,C,D	A,C,D	C,D	C,D	C,D	A,C,D	A,C,D	A,C,D	C,D	M	
Jackson	Yes	1,4	A,C,D	C,D	A,C,D	A,C,D	A,C,D	A,C,D	C,D	C,D	C,D	C,D	C,D	C,D	C,D	S	
Misssouri	No																
Kansas City	Yes	4	A,C,D	A,C,D	C,D	C,D	C,D	C,D	C,D	C,D	C,D			C,D	C,D	M,Q	A
Montana	Yes	2,3,4	A,C,D	A,C,D	C,D	C,D	C,D	C,D	C,D	C,D	C,D			C,D	C,D	M,Q	A
Nebraska	Yes	1,4	A,C,D	A,C,D	C,D	C,D	C,D	C,D	C,D	C,D					C,D	M	

NOTE: An explanation of the numbers and letters used is given at the end of the table (see p 42).

Table 1 *(Continued)*

State/City	Does State/City Have Legislation?	Legislation Code Type?	Hospitals	Nursing Homes	Schools	Theaters (Public Gathering Places)	Office Buildings	Hotels	Apartment Buildings	Airports	Fire and Police Stations	Water Treatment Plants	Sewage Treatment Plants	All Public Buildings, State	All Public Buildings, Commercial	Applicable Government Agency	Smokeproof Enclosures in High-Rise Buildings
Lincoln	Yes	4,8	A,C,D	A,C,D	C,D	C,D	C,D	C,D	C,D	C,D					C,D	M,S	
Omaha	Yes	4,8	A,C,D	A,C,D		C,D	C,D	C,D	C,D	C,D					C,D	S	
Nevada	Yes	1,2,3,4	A,C,D	A,C,D	C,D	C,D	C,D	C,D	C,D	C,D				A,C,D	C,D	M,O,Q	A*
New Hampshire	Yes	1,2,4	A,C,D	A,C,D	C,D	C,D	C,D	C,D	C,D	C,D				C,D	C,D	M	A
New Jersey	Yes	2,3,4,5	A,C,D	A,C,D	C,D	C,D	C,D	C,D	C,D	C,D	C,D	C,D	C,D	C,D	C,D	Q	
New Mexico	Yes	1,2,3,4	A,C,D	A,C,D	A,C,D	A,C,D	A,C,D	A,C,D	C,D	A,C,D	C,D	C,D	C,D	C,D	A,C,D	Q	
Albuquerque	Yes	3,4	A,C,D	A,C,D	C,D	C,D	C,D	C,D	C,D	C,D				C,D	C,D	S	A
New York	Yes	2,4	A,C,D	A,C,D	C,D	C,D	C,D	C,D	C,D	C,D				C,D	C,D	O	
Albany	Yes	2,4	A,C,D	A,C,D	C,D	C,D	C,D	C,D	C,D	C,D		B,C,D	B,C,D	C,D	C,D	Q	
Buffalo	Yes	2,8	A,C,D	A,C,D	C,D	C,D	C,D	C,D	C,D	C,D		B,C,D	B,C,D	C,D	C,D	N,O,R	
New York	Yes	2,8	A,C,D	A,C,D	C,D	C,D	C,D	C,D	C,D	C,D	C,D	B,C,D	B,C,D	C,D	C,D	Q	B
Syracuse	Yes	3,8	A,C,D	A,C,D	C,D	C,D	C,D	C,D	C,D	C,D				C,D	C,D	M,Q	A
North Carolina	Yes	2,4,6	A,C,D	A,C,D	C,D	C,D	C,D	C,D	C,D	C,D	C,D	C,D	C,D	C,D	C,D	O,T	A
North Dakota	Yes	1,3,4	A,C,D	A,C,D	C,D	C,D	C,D	C,D	C,D	C,D	C,D	C,D	C,D	C,D	C,D	M,O	A
Ohio	Yes	2,4,5	A,C,D	A,C,D	C,D	C,D	C,D	C,D	C,D	C,D	C,D	C,D	C,D	C,D	C,D	Q,S	A
Akron	Yes	2,4,5	A,C,D	A,C,D	C,D	A,C,D	A,C,D	A,C,D	A,C,D	C,D	C,D	C,D	C,D	C,D	C,D	O	A
Cincinnati	Yes	2,4,5	A,C,D	A,C,D	C,D	A,C,D	A,C,D	A,C,D	A,C,D	C,D	C,D	C,D	C,D	C,D	C,D	Q,S	A
Cleveland	Yes	2,4,5	A,C,D	A,C,D	C,D	C,D	C,D	C,D	C,D	C,D	C,D	C,D	C,D	C,D	C,D	M,Q	A
Dayton	Yes	2,4	A,C,D	A,C,D	C,D	C,D	C,D	C,D	C,D	C,D	C,D	C,D	C,D	C,D	C,D	Q	
Youngstown	Yes†	2,4,5	A,C,D	A,C,D	C,D	C,D	C,D	C,D	C,D	C,D	C,D	C,D	C,D	C,D	C,D	Q	
Oklahoma	Yes	1,2	A,C,D	C,D	C,D	C,D	C,D	C,D	C,D	A,C,D	A,C,D			C,D	C,D	M	
Oregon	Yes	1,2,3,4	A,C,D	A,C,D	C,D	C,D	C,D	A,C,D	A,C,D	C,D		C,D	C,D	C,D	C,D	M,Q	B
Portland **	Yes	1,2,4	A,C,D	C,D	C,D	C,D	C,D	A,C,D	A,C,D	A,C,D	A,C,D			A,C,D	C,D	M,Q	
Pennsylvania	Yes	1,2	A,C,D	A,C,D	A,C,D	A,C,D	A,C,D	A,C,D	A,C,D	A,C,D				A,C,D	A,C,D	R	A
Philadelphia	Yes	4,8	A,C,D	A,C,D	A,C,D	A,C,D	A,C,D	A,C,D	A,C,D	A,C,D	C,D	C,D	C,D	A,C,D	A,C,D	M,Q	A
Rhode Island	Yes	2,4,5	A,C,D	A,C,D	C,D	C,D	C,D	C,D	C,D	C,D	C,D	C,D	C,D	C,D	C,D	M,Q	
South Carolina	Yes	1,4,6	A,C,D	A,C,D	C,D	C,D	C,D	C,D	C,D	C,D	C,D			C,D	C,D	M,O	A

NOTE: An explanation of the numbers and letters used is given at the end of the table (see p 42).
† State buildings only.

Table 1 (Continued)

State/City	Does State/City Have Legislation?	Legislation Code Type?	Hospitals	Nursing Homes	Schools	Theaters (Public Gathering Places)	Office Buildings	Hotels	Apartment Buildings	Airports	Fire and Police Stations	Water Treatment Plants	Sewage Treatment Plants	All Public Buildings, State	All Public Buildings, Commercial	Applicable Government Agency	Smokeproof Enclosures in High-Rise Buildings
South Dakota	Yes	1,2,4	A,C,D	A,C,D	C,D	C,D	C,D	C,D	C,D	C,D				C,D	C,D	M	A
Tenessee	Yes	1,4,6	A,C,D	A,C,D	C,D	C,D	C,D	C,D	C,D					C,D	C,D	M,N,O	
Texas	Yes	2	A,C,D	B,C,D												M,O,U	
Amarillo **	Yes	8	A,C	A,C	C,D	C,D	C,D	C,D	C	A,C	A,C	A	A	A	C	O	
Austin	Yes	3	A,C,D	A,C,D	C,D	C,D	C,D	C,D	C,D	C,D	C,D			C,D	C,D	Q	
Corpus Christi **	Yes	3	A,C,D	A,C,D	C,D	A,C,D	C,D	C,D	C,D	A,C,D				C,D	C,D	Q	
Dallas **	Yes	3,4	A,C,D	A,C,D	C,D	A,C,D	A,C,D	A,C,D	A,C,D	C,D	C,D			A,C,D	A,C,D	S	A
El Paso	Yes	4,6	A,C,D	A,C,D	C,D	C,D	C,D	C,D	C,D		C,D	C,D		C,D	C,D	S	
Fort Worth **	Yes	3,4	A,C,D	C,D	C,D	C,D	C,D	C,D	C,D	C,D	A,C,D			C,D	C,D	S	
Houston	Yes	3,4,8	A,C,D	A,C,D	C,D	C,D	B,C,D*	B,C,D*	B,C,D*	B,C,D*	B,C,D*			B,C,D*	B,C,D*	S	B
Lubbock	Yes	1,3,4	A,C,D	A,C,D	C,D	C,D	C,D	C,D	C,D	C,D	C,D			C,D	C,D	M,Q	
San Antonio **	Yes	4,8	A,C,D	A,C,D	C,D	A,C,D	C,D	C,D	C,D	A,C,D	C,D			A,C,D	A,C,D	M	B
Wichita Falls **	Yes	7	A,C,D	A,C,D	C,D	C,D	C,D	C,D	C,D	C,D	C,D			C,D	C,D	S	
Utah	Yes	1,2,3,4	A,C,D	A,C,D	C,D	C,D	C,D	C,D	C,D	C,D	A,C,D	C,D	C,D	C,D	C,D	M,Q	A
Salt Lake City **	Yes	3,8	C,D	C,D	C,D	C,D	C,D	C,D	C,D	C,D	C,D			C,D	C,D	M,Q	B
Vermont	Yes	1,2,4,5	A,C,D	A,C,D	C,D	C,D	C,D	C,D	C,D	C,D	C,D			C,D	C,D	R	
Virginia	Yes	2,4,5	A,C,D	C,D	C,D	C,D	C,D	C,D	C,D	C,D	C,D	C,D	C,D	C,D	C,D	O	
Richmond	Yes	1,4,5	A,C,D	C,D	C,D	C,D	C,D	C,D	C,D	C,D	C,D			C,D	C,D	O,Q	
Virginia Beach	Yes	4,5	A,C,D	A,C,D	A,C,D	A,C,D	C,D	C,D	C,D	A,C,D	C,D			C,D	C,D	Q	
Washington	Yes	2,3,4	A,C,D	A,C,D	C,D	C,D	C,D	C,D	C,D	C,D	B,C,D	A,B	A,B	A,B	C,D	R	
Seattle **	Yes	3,4	A,C,D	A,C,D	C,D	C,D	C,D	A,C,D	A,C,D	C,D	C,D			A,B	C,D	Q	B
West Virginia	Yes	1,2,4	A,C,D	A,C,D	C,D	C,D	C,D	C,D	A,C,D	C,D	C,D			C,D	C,D	M,O	
Wisconsin	Yes	2,4	A,C,D	A,C,D	C,D	C,D	C,D	C,D	C,D	C,D	C,D			C,D	C,D	R,S	
Madison	Yes	2,4	A,C,D	A,C,D	C,D	C,D	C,D	C,D	C,D	C,D	C,D			C,D	C,D	O	
Milwaukee	Yes	2,4,8	A,C,D	A,C,D	C,D	C,D	C,D	C,D	C,D	C,D	C,D			C,D	C,D	Q	
Wyoming	Yes	1,2,3	A,C,D	A,C,D	A,C,D	C,D	C,D	C,D	C,D	C,D	C,D			C,D	C,D	M,N,P	

NOTE: An explanation of the numbers and letters used is given at the end of the table (see p 42).

Table 1 *(Continued)*

Explanation of Numbers and Letters Used in Table 1:

Legislation Code Type

1. Life Safety Code, ANSI/NFPA 101-1985 [11]
2. State
3. Uniform Building Code [24]
4. National Electrical Code, ANSI/NFPA 70-1987 [9]
5. Building Officials and Code Administration (BOCA)
6. Standard Building Code [23]
7. Health Care Facilities Code, ANSI/NFPA 99-1984 [10]
8. City

Power Source

A. Emergency Power
B. Standby Power
C. Exit Lighting
D. Egress Lighting

Governing Agency

M. Fire Marshal or Division of Fire
N. Department of Public Health
O. Local Government Units
P. Public Safety
Q. Building Commission or Department
R. Department of Labor
S. Inspection Department
T. Department of Insurance
U. Various, but usually depends on occupancy

*High-rise building.
**No changes made since previous report.

Table 2 lists the needs in thirteen general categories with some breakdown under each to indicate major requirements. Ranges under the columns "Maximum Tolerance Duration of Power Failure" and "Recommended Minimum Auxiliary Supply Time" are assigned based upon experience. Written standards have been referenced where applicable.

In some cases, under the column "Type of Auxiliary Power System" both emergency and standby have been indicated as required. An emergency supply of limited time capacity may be used at a low cost for immediate or uninterruptible power until a standby supply can be brought on line. An example would be the case where battery lighting units come on until a standby generator can be started and transferred to the critical loads.

Following the table, the "General Need" listings are presented in greater detail with recommendations as to the type of equipment or system that should be used.

Users of this standard may wish to skip the detailed presentation of each "General Need" and go directly to Chapter 4, "Systems and Hardware." If so, care should be taken that all individual needs have been recognized and listed so that suitable power systems can be selected to meet all requirements.

Readers using this standard may find that various combinations of general needs will require an in-depth system and cost analysis that will modify the recommended equipment and systems to best meet the specific requirements.

Small commercial establishments and manufacturing plants will usually find their requirements under two or three of the general need guidelines given in this chapter. Large manufacturers and commercial facilities will find that portions or all of the need guidelines given here apply to their operations and justify or require emergency and backup standby electric power.

3.2 Lighting

3.2.1 Introduction. Evaluation of the quality, quantity, type, and duration of emergency or standby power for lighting is necessary for each particular application. The different types of systems have various degrees of reliability that must be considered in the selection of the proper system.

3.2.2 Lighting for Evacuation Purposes. Interruption of power to a normal lighting system may cause injury or loss of life. Emergency lighting for evacuation purposes must energize automatically upon loss of normal lighting. Where legally required (ANSI/NFPA 101-1985 [11]) emergency systems are installed, lighting must be maintained for at least 1½ h if battery-powered unit equipment is used. Emergency lighting must provide enough illumination to allow easy and safe egress from the area involved. All exit lights, signs, and stairwell lights should be included in both the emergency lighting system and the normal lighting system. Design of the emergency lighting system should include consideration of the need for lighting to silhouette protruding machines or objects in aisles.

3.2.3 Perimeter and Security Lighting. Emergency or standby power for perimeter and security lighting may be deemed necessary to reduce risk of injury, theft, or property damage. The power for perimeter lighting may not be required until several minutes after failure of normal power. In order to maintain perimeter lighting throughout the dark hours, a system should be capable of supplying power

Table 2
Condensed General Criteria for Preliminary Consideration

Section	General Need	Specific Need	Maximum Tolerance Duration of Power Failure	Recommended Minimum Auxiliary Supply Time	Type of Auxiliary Power System		System Justification
					Emergency	Standby	
3.2	Lighting	Evacuation of personnel	Up to 10 s, preferably not more than 3 s	2 h	×		Prevention of panic, injury, loss of life; Compliance with building codes and local, state, and federal laws; Lower insurance rates; Prevention of property damage; Lessening of losses due to legal suits
		Perimeter and security	10 s	10–12 h during all dark hours	×	×	Lower losses from theft and property damage; Lower insurance rates; Prevention of injury
		Warning	From 10 s up to 2 or 3 min	To return to prime power source	×		Prevention or reduction of property loss; Compliance with building codes and local, state, and federal laws; Prevention of injury and loss of life
		Restoration of normal power system	1 s to indefinite depending on available light	Until repairs completed and power restored	×	×	Risk of extended power and light outage due to a longer repair time
		General lighting	Indefinite; depends on analysis and evaluation	Indefinite; depends on analysis and evaluation		×	Prevention of loss of sales; Reduction of production losses; Lower risk of theft; Lower insurance rates

Table 2 *(Continued)*

Section	General Need	Specific Need	Maximum Tolerance Duration of Power Failure	Recommended Minimum Auxiliary Supply Time	Type of Auxiliary Power System		System Justification
					Emergency	Standby	
		Hospitals and medical areas	Uninterruptible to 10 s ANSI/NFPA 99-1984 [10], 101-1985 [11] allow 10 s for alternate power source to start and transfer	To return of prime power	×	×	Facilitate continuous patient care by surgeons, medical doctors, nurses, and aids; Compliance with all codes, standards, and laws; Prevention of injury or loss of life; Lessening of losses due to legal suits
		Orderly shutdown time	0.1 s to 1 h	10 min to several hours	×		Prevention of injury or loss of life; Prevention of property loss by a more orderly and rapid shutdown of critical systems; Lower risk of theft; Lower insurance rates
3.3	Startup power	Boilers	3 s	To return of prime power	×		Return to production; Prevention of property damage due to freezing; Provision of required electric power
		Air compressors	1 min	To return of prime power		×	Return to production; Provision for instrument control
3.4	Transportation	Elevators	15 s to 1 min	1 h to return of prime power		×	Personnel safety; Building evacuation; Continuation of normal activity
		Material handling	15 s to 1 min	1 h to return of prime power		×	Completion of production run; Orderly shutdown; Continuation of normal activity
		Escalators	15 s to no requirement for power	Zero to return of prime power		×	Orderly evacuation; Continuation of normal activity

Table 2 *(Continued)*

Section	General Need	Specific Need	Maximum Tolerance Duration of Power Failure	Recommended Minimum Auxiliary Supply Time	Type of Auxiliary Power System		System Justification
					Emergency	Standby	
		Conveyors	15 s to 1 min	As analyzed and economically justified		×	Completion of production run Completion of customer order Orderly shutdown Continuation of normal activity
3.5	Mechanical utility systems	Water (cooling and general use)	15 s	½ h to return of prime power		×	Continuation of production Prevention of damage to equipment Supply of fire protection
		Water (drinking and sanitary)	1 min to no requirement	Indefinite until evaluated		×	Providing of customer service Maintaining personnel performance
		Boiler power	0.1 s	1 h to return of prime power	×	×	Prevention of loss of electric generation and steam Maintaining production Prevention of damage to equipment
		Pumps for water, sanitation, and production fluids	10 s to no requirement	Indefinite until evaluated		×	Prevention of flooding Maintaining cooling facilities Providing sanitary needs Continuation of production Maintaining boiler operation
		Fans and blowers for ventilation and heating	0.1 s to return of normal power	Indefinite until evaluated	×	×	Maintaining boiler operation Providing for gas-fired unit venting and purging Maintaining cooling and heating functions for buildings and production
3.6	Heating	Food preparation	5 min	To return of prime power		×	Prevention of loss of sales and profit Prevention of spoilage of in-process preparation

Table 2 *(Continued)*

Section	General Need	Specific Need	Maximum Tolerance Duration of Power Failure	Recommended Minimum Auxiliary Supply Time	Type of Auxiliary Power System		System Justification
					Emergency	Standby	
		Process	5 min	Indefinite until evaluated; normally for time for orderly shutdown, or to return of prime power		×	Prevention of in-process product damage; Prevention of property damage; Continued production; Prevention of payment to workers during no production; Lower insurance rates
3.7	Refrigeration	Special equipment or devices which have critical warmup (cryogenics)	5 min	To return of prime power		×	Prevention of equipment or product damage
		Depositories of critical nature (blood banks, etc)	5 min (10 s per ANSI/NFPA 99-1984 [10]	To return of prime power		×	Prevention of loss of material stored
		Depositories of noncritical nature (meat, produce, etc)	2 h	Indefinite until evaluated		×	Prevention of loss of material stored; Lower insurance rates
3.8	Production	Critical process power (sugar factory, steel mills, chemical processes, glass products, etc)	1 min	To return of prime power or until orderly shutdown		×	Prevention of product and equipment damage; Continued normal production; Reduction of payment to workers on guaranteed wages during nonproductive period; Lower insurance rates; Prevention of prolonged shutdown due to nonorderly shutdown

Table 2 *(Continued)*

Section	General Need	Specific Need	Maximum Tolerance Duration of Power Failure	Recommended Minimum Auxiliary Supply Time	Type of Auxiliary Power System — Emergency	Type of Auxiliary Power System — Standby	System Justification
		Process control power	Uninterruptible (UPS) to 1 min	To return of prime power	×	×	Prevention of loss of machine and process computer control program Maintaining production Prevention of safety hazards from developing Prevention of out-of-tolerance products
3.9	Space conditioning	Temperature (critical application)	10 s	1 min to return of prime power	×	×	Prevention of personnel hazards Prevention of product or property damage Lower insurance rates Continuation of normal activities Prevention of loss of computer function
		Pressure (critical) pos/neg atmosphere	1 min	1 min to return of prime power	×	×	Prevention of personnel hazards Continuation of normal activities Prevention of product or property damage Lower insurance rates Compliance with local, state, and federal codes, standards, and laws
		Humidity (critical)	1 min	To return of prime power	×	×	Prevention of loss of computer functions Maintenance of normal operations and tests Prevention of explosions or other hazards

Table 2 (*Continued*)

Section	General Need	Specified Need	Maximum Tolerance Duration of Power Failure	Recommended Minimum Auxiliary Supply Time	Type of Auxiliary Power System		System Justification
					Emergency	Standby	
		Static charge	10 s or less	To return of prime power	×		Prevention of static electric charge and associated hazards Continuation of normal production (printing press operation, painting spray operations)
		Building heating and cooling	30 min	To return of prime power	×		Prevention of loss due to freezing Maintenance of personnel efficiency Continuation of normal activities
		Ventilation (toxic fumes)	15 s	To return of prime power or orderly shutdown	×		Reduction of health hazards Compliance with local, state, and federal codes, standards, and laws Reduction of pollution
		Ventilation (explosive atmosphere)	10 s	To return of prime power or orderly shutdown	×		Reduction of explosion hazard Prevention of property damage Lower insurance rates Compliance with local, state, and federal codes, standards, and laws Lower hazard of fire Reduce hazards to personnel
		Ventilation (building general)	1 min	To return of prime power	×		Maintaining of personnel efficiency Providing make-up air in building

Table 2 *(Continued)*

Section	General Need	Specified Need	Maximum Tolerance Duration of Power Failure	Recommended Minimum Auxiliary Supply Time	Type of Auxiliary Power System		System Justification
					Emergency	Standby	
		Ventilation (special equipment)	15 s	To return of prime power or orderly shutdown	X	X	Purging operation to provide safe shutdown or startup; Lowering of hazards to personnel and property; Meeting requirements of insurance company; Compliance with local, state, and federal codes, standards, and laws; Continuation of normal operation
		Ventilation (all categories noncritical)	1 min	Optional		X	Maintaining comfort; Preventing loss of tests
		Air pollution control	1 min	Indefinite until evaluated; compliance or shutdowns are options	X	X	Continuation of normal operation; Compliance with local, state, and federal codes, standards, and laws
3.10	Fire protection	Annunciator alarms	1 s	To return of prime power	X		Compliance with local, state, and federal codes, standards, and laws; Lower insurance rates; Minimizing life and property damage
		Fire pumps	10 s	To return of prime power		X	Compliance with local, state, and federal codes, standards, and laws; Lower insurance rates; Minimizing life and property damage

Table 2 *(Continued)*

Section	General Need	Specified Need	Maximum Tolerance Duration of Power Failure	Recommended Minimum Auxiliary Supply Time	Type of Auxiliary Power System Emergency	Type of Auxiliary Power System Standby	System Justification
		Auxiliary lighting	10 s	5 min to return of prime power	×	×	Servicing of fire pump engine should it fail to start Providing visual guidance for fire-fighting personnel
3.11	Data processing	CPU memory tape/disk storage, peripherals	½ cycle	To return of prime power or orderly shutdown	×	×	Prevention of program loss Maintaining normal operations for payroll, process control, machine control, warehousing, reservations, etc
		Humidity and temperature control	5 to 15 min (1 min for water-cooled equipment)	To return of prime power or orderly shutdown	×	×	Maintenance of conditions to prevent malfunctions in data processing system Prevention of damage to equipment Continuation of normal activity
3.12	Life support and life safety systems (medical field, hospitals, clinics, etc)	X-ray	Milliseconds to several hours	From no requirement to return of prime power, as evaluated	×	×	Maintenance of exposure quality Availability for emergencies
		Light	Milliseconds to several hours	To return of prime power	×	×	Compliance with local, state, and federal codes, standards, and laws Preventing interruption to operation and operating needs
		Critical to life, machines, and services	½ cycle to 10 s	To return of prime power	×	×	Maintenance of life Prevention of interruption of treatment or surgery Continuation of normal activity Compliance with local, state, and federal codes, standards, and laws

51

Table 2 (*Continued*)

Section	General Need	Specified Need	Maximum Tolerance Duration of Power Failure	Recommended Minimum Auxiliary Supply Time	Type of Auxiliary Power System		System Justification
					Emergency	Standby	
		Refrigeration	5 min	To return of prime power		×	Maintaining blood, plasma, and related stored material at recommended temperature and in prime condition
3.13	Communication systems	Teletypewriter	5 min	To return of prime power		×	Maintenance of customer services Maintenance of production control and warehousing Continuation of normal communication to prevent economic loss
		Inner building telephone	10 s	To return of prime power	×		Continuation of normal activity and control
		Television (closed circuit and commercial)	10 s	To return of prime power		×	Continuation of sales Meeting of contracts Maintenance of security Continuation of production
		Radio systems	10 s	To return of prime power	×	×	Maintenance of security and fire alarms Providing evacuation instructions Continuation of service to customers Prevention of economic loss Directing vehicles normally
		Intercommunication systems	10 s	To return of prime power	×	×	Providing evacuation instructions Directing activities during emergency Providing for continuation of normal activities Maintaining security

ANSI/IEEE
Std 446-1987

Table 2 (*Continued*)

Section	General Need	Specific Need	Maximum Tolerance Duration of Power Failure	Recommended Minimum Auxiliary Supply Time	Type of Auxiliary Power System		System Justification
					Emergency	Standby	
		Paging systems	10 s	½ h	×	×	Locating of responsible persons concerned with power outage Providing evacuation instructions Prevention of panic
3.14	Signal circuits	Alarms and annunciation	1 to 10 s	To return of prime power	×	×	Prevention of loss from theft, arson, or riot Maintaining security systems Compliance with codes, standards, and laws Lower insurance rates Alarm for critical out-of-tolerance temperature, pressure, water level, and other hazardous or dangerous conditions Prevention of economic loss
		Land-based aircraft, railroad, and ship warning systems	1 s to 1 min	To return of prime power	×	×	Compliance with local, state, and federal codes, standards, and laws Prevention of personnel injury Prevention of property and economic loss

for 10–12 h for every 24 h the normal power source is off. For this reason, the unit battery equipment is not recommended for auxiliary perimeter lighting.

3.2.4 Warning Lights. Emergency power must be available for all warning lights, such as aircraft warning lights on high structures, ship warning lights on edges of waterways, and other warning lights that act to prevent injury or property damage. The power source selected should be capable of supplying emergency power throughout the duration of the longest anticipated power failure; therefore, the unit battery type is normally not suited to this application.

3.2.5 Health Care Facilities. Emergency lighting is of paramount importance in hospitals and similar institutions. The requirements for these areas are included in 3.12.

3.2.6 Standby Lighting for Equipment Repair. Standby power for lighting should be installed in areas where the most probable internal power system failures may occur and in the main switchgear rooms. This requirement is justified by the necessity of having enough light to repair the equipment that failed and caused the loss of normal lighting.

3.2.7 Lighting for Production. Interruption of power to a normal lighting system may cause serious curtailment or complete loss of production. Where there are no safety hazards or property damage associated with this need, the decision should be based on the economic evaluation of each particular application. Systems that provide power for emergency lighting may also provide high-level lighting to allow production to continue.

3.2.8 Lighting to Reduce Hazards to Machine Operators. A machine operator may be subjected to a high injury risk for the first few seconds after lighting has failed. Many machines present a safety hazard if suddenly plunged into darkness. Instantaneous emergency lighting is required for protection against this type of injury.

3.2.9 Supplemental Lighting for High-Voltage Discharge Systems. If mercury or other types of high-voltage discharge lighting are used for the regular system, consideration should be given to adding auxiliary lamps such as incandescent or fluorescent. Some high-voltage discharge lamps require a cooling period before they restrike the arc and a warm-up period before they attain full brilliance. The total time required for full illumination after a momentary power interruption can range from 1 min for high pressure sodium to 20 min for metal halide and mercury vapor.

3.2.10 Codes, Rules, and Regulations. Many states and municipalities have adopted their own specific codes regarding emergency lighting, in addition to those set down by the following organizations:

(1) Occupational Safety and Health Act of 1970 (OSHA) [25]; OSHA is charged with enforcing compliance and makes reference to the NEC and other NFPA codes

(2) ANSI/NFPA 70-1987 [9], the National Electrical Code (NEC), Article 700, sets forth the standard of practice for emergency lighting equipment with regard to installation, operation, and maintenance

(3) ANSI/NFPA 101-1985 [11], the Life Safety Code, concerns itself with the specification of locations where emergency lighting is considered essential to life safety and specifics on exit marking

(4) Underwriter's Laboratories, Inc, tests and approves equipment to uniform performance standards as established by ANSI/UL 924-1983 [13]

3.2.11 Recommended Systems. For short time durations, primarily lighting for personnel safety and evacuation purposes, battery units are satisfactory. Where longer service and heavier loads are required, an engine- or turbine-driven generator is usually used, which starts automatically upon failure of the prime power source with the load applied by an automatic transfer switch [17]. It is generally considered that an average level of 0.4 footcandles is adequate where passage is required and no precise operations are expected [21].

Table 3 summarizes the user's needs for emergency and standby electric power for lighting by application and areas.

Table 3
Typical Emergency and Standby Lighting Recommendations

Standby*	Immediate, Short-Term†	Immediate, Long-Term‡
Security lighting	Evacuation lighting	Hazardous areas
Outdoor perimeters	Exit signs	Laboratories
Closed circuit TV	Exit lights	Warning lights
Night lights	Stairwells	Storage areas
Guard stations	Open areas	Process areas
Entrance gates	Tunnels	
	Halls	Warning lights
Production lighting		Beacons
Machine areas	Miscellaneous	Hazardous areas
Raw materials storage	Standby generator areas	Traffic signals
Packaging	Hazardous machines	
Inspection		Health care facilities
Warehousing		Operating rooms
Offices		Delivery rooms
		Intensive care areas
Commercial lighting		Emergency treatment areas
Displays		
Product shelves		Miscellaneous
Sales counters		Switchgear rooms
Offices		Elevators
		Boiler rooms
Miscellaneous		Control rooms
Switchgear rooms		
Landscape lighting		
Boiler rooms		
Computer rooms		

* An example of a standby lighting system is an engine-driven generator.
† An example of an immediate short-term lighting system is the common unit battery equipment.
‡ An example of an immediate long-term lighting system is a central battery bank rated to handle the required lighting load only until a standby engine-driven generator is placed on-line.

3.3 Startup Power

3.3.1 Introduction. Assume a *cold* boiler and a *dead* plant without electric power or steam. From this premise, several very important issues should be considered, such as:

(1) How will the plant be protected from freezing in cold weather? Even with gas heaters, will there be sufficient heat without fans and without interlocked make-up air units running?

(2) A steam turbine generator is on hand, but without forced draft, induced draft, boiler feed water, flame detectors, or control power. How can it be started?

(3) A gas turbine generator has been installed, but how can this be started without bringing it up in speed with a small steam turbine, an electric motor, or other prime mover? Gas compressors may be necessary and also require prime movers of some type.

(4) Steam- and electrically driven fire pumps are out of service. There may be no major fire protection until electric power or steam is restored.

(5) An uninterruptible power supply of sufficient capacity is probably not on hand; otherwise, steam and electric power would not be down.

These statements illustrate the fact that adequate startup power is one of the most important considerations in the original design of any plant. Millions of dollars' worth of equipment could be standing idle at a time of critical need if no allowance were made for starting the machines under unexpected conditions such as a major power outage.

3.3.2 Example of a System Utilizing Startup Power. Starting major plant equipment without outside power is commonly referred to as a *black start* and is accomplished by using only the facilities available within the plant. One example of a system designed with *black start* capability, with a minimum electrical startup system, would be a large gas turbine driving a centrifugal compressor in the natural gas pipeline industry, where the high-pressure gas from the pipeline is used to drive expansion turbines and gas motors for cranking the turbine, operating pumps, and positioning valves. By utilizing the high-pressure gas for the large horsepower requirements, a small engine-driven generator, fueled by natural gas, is used to provide electric power for turbine accessories, battery charging, lighting, and powering other critical loads. When the turbine is running, a shaft-driven generator provides larger quantities of electric power for all station requirements; the small generator is placed on *standby*.

3.3.3 Lighting. In the design of the startup power system, first consideration should be given to installing battery-operated lights in the vicinity of the standby power source and switchgear.

3.3.4 Engine-Driven Generators. The battery-cranking power for the engine may also be used for some lights. Cranking may also be accomplished by compressed air supplied to a tank by the plant compressed-air system and prevented from leaving the tank into the plant system by a check valve.

The engines may be sized for short-term operation if some type of continuous power generation can be brought onto the line after startup. If no generation as a prime source of electric power has been installed, the diesel or engine-generator should be sized for supporting all electrical needs for the generator auxiliaries, boilers, critical emergency lights, fire signals, exit lights, and other items listed in Table 2.

3.3.5 Battery Systems. Special consideration should be given to the design of the plant battery system or uninterruptible power supply, allowing for (1) adequate battery capacity to provide power for the necessary startup control systems, following a programmed safe stop; and (2) special disconnecting devices to automatically disconnect large power-consuming systems from the battery system, when possible, to conserve battery capacity for restart.

With capacity for these minimum facilities in operation, consideration may then be given to installing sufficient capacity to support additional justified needs.

3.3.6 Other Systems. Mobile equipment may suffice if it can reasonably be assumed to be available when needed. (Who has the highest priority when all have the need?) An alternate standby public utility line may also be available at a low cost from a separate source of supply. A neighboring plant with live generation may assist in an emergency.

3.3.7 System Justification. A definite workable plan with the proper equipment should be evaluated, and action taken as justified, prior to the need.

3.4 Transportation

3.4.1 Introduction. This topic covers the moving of people and products by methods that depend upon electric power and the importance of maintaining power ranges from desirable to critical.

3.4.2 Elevators. Where two or more elevators are in use in buildings three or more stories high, the elevators or banks of elevators should be connected to separate sources of power. There are situations where standby power is required for all elevators within 15 s. Savings may be made by supplying power during outages of the normal supply to half the elevators installed, providing the traffic can be rerouted and the capacity of the elevators is adequate. Power must be transferred to the second bank of elevators within 1 min or so of the prime power loss to clear stalled elevators. Power may be left on this bank until normal power returns.

Where elevator service is critical for personnel and patients, it is desirable to have automatic power transfer with manual supervision. Operators and maintenance men may not be available in time if the power failure occurs on a weekend or at night.

(1) *Typical Elevator System.* Figure 3 shows an elevator emergency power transfer system whereby one preferred elevator is fed from a vital load bus through an emergency riser, while the rest of the elevators are fed from the normal service. By providing an automatic transfer switch for each elevator and a remote selector station, it is possible to select individual elevators, thus permitting complete evacuation in the event of power failure. The engine-generator set and emergency riser need only be sized for one elevator, thus minimizing the installation cost. The controls for the remote selector, automatic transfer switches, and engine starting are independent of the elevator controls, thereby simplifying installation.

(2) *Regenerated Power.* Regenerated power is a concern for motor-generator-type elevator applications. In some elevator applications, the motor is used as a

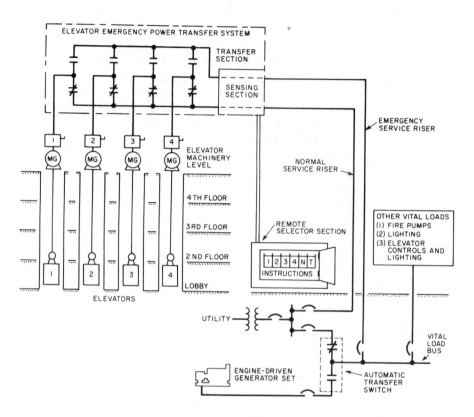

Fig 3
Elevator Emergency Power Transfer System

brake when the elevator is descending and generates electricity. Electric power is then pumped back into the power source. If the source is commercial utility power, it can easily be absorbed. If the power source is an engine-driven generator, the regenerated power can cause the generating set and the elevator to overspeed. To prevent overspeeding of the elevator, the maximum amount of power that can be pumped back into the generating set must be known. The permissible amount of absorption is approximately 20% of the generating set's rating in kilowatts. If the amount pumped back is greater than 20%, other loads must be connected to the generating set, such as emergency lights or "dummy" load resistances. Emergency lighting should be permanently connected to the generating set for maximum safety. A dummy load can also be automatically switched on the line whenever the elevator is operating from an engine-driven generator.

3.4.3 Conveyors and Escalators. Escalators and personnel conveyors may require emergency power since physically handicapped persons ride up and would have great difficulty walking down, even though a normal person would be able to do so if the power were off.

Those who have used conveyors or elevators on a small scale over a period of time and have continued to grow should be wary and check their needs for standby power. To feed a few cattle or gather a few eggs, using conveyors may have once eased someone's labor, with labor filling in if needed. As operations become larger and larger, there is a point at which an outage of electric power could produce a disaster. The gathering of 250 000 eggs a day or the feeding of 10 000 head of cattle is dependent upon power when needed.

3.4.4 Other Transportation Systems. Power for charging equipment for battery-powered vehicles is usually a noncritical requirement. Time available for the emergency generation to come on line varies from several minutes to several hours.

A few examples of transportation systems that may need standby power are

(1) Conveyors for raw materials through to finished goods
(2) Warehouse high-stacking loading and unloading equipment and conveyors delivering finished goods to shipping facilities
(3) Slurry pumps for long pipe lines
(4) Livestock feeder conveyors

3.5 Mechanical Utility Systems

3.5.1 Introduction. Often the need for mechanical utilities is as great as that for electric power. These are interdependent. We can speak of the utility systems as a whole since, to most managers and corporate officials, these are all united in a group under the heading "utilities."

3.5.2 Typical Utility Systems for Which Reliable Power May Be Necessary. Mechanical utility systems comprise the following services for which reliable electric power is usually required:

(1) Compressed air for pneumatic power
(2) Cooling water (including return pumps, pressure pumps, tower fans)
(3) Well water or other pumped sources for personnel use
(4) Hydraulic systems (200 lb, 1500 lb, or other pressures as required)
(5) Sewer systems (sanitary, industrial, storm)
(6) Gas systems (natural, propane, oxygen), including compressors
(7) Fire pumps and associated water supplies
(8) Steam systems (low and high pressures)
(9) Ventilation (building and process)
(10) Vacuum systems
(11) Compressed air for instrumentation

Additional systems may exist in some plants, but the list will alert a plant manager or engineer to the various needs and potential losses should electric power not be available.

Systems may be required for manufacturing and services to maintain other services. For example, electric power, water, and compressed air for boilers used to supply steam for the generation of electricity.

A 0.1–5 s power failure may cause operating engineers to spend minutes or hours restarting equipment and making adjustments until all systems are again stable. Such disruptions to production should be prevented if economically justifiable.

3.5.3 Orderly Shutdown of Mechanical Utility Systems. An orderly shutdown may be acceptable and can be provided with a short-time smaller supply source of power for the following requirements:

(1) Maintain temperature or pressure on vulcanizers until the product can be finished.

(2) Maintain hydraulic pressure until a batch process is completed, or until the pressure can be released without loss.

(3) Operate pumps for a time until all process water has been shut off or has drained back into the sumps to prevent flooding. This becomes serious when the water is contaminated with oil and other waste when flooding occurs and special handling may be needed to comply with antipollution requirements.

(4) Maintain ventilation to clear explosive atmospheres while a normal shutdown proceeds. Purging air is critical to some oven-drying processes.

(5) Prevent sanitary sewers from overflowing before personnel evacuation takes place.

(6) Run gas compressors for the finishing of a critical process.

A complete listing has not been intended, but representative needs of various types should be enough to alert a plant engineer so that a plan may be prepared for an orderly shutdown. Management may then act on the needs justified without being caught unprepared in an emergency situation. Although management may not approve the recommendations immediately, the plan will be ready for resubmitting, approval, and implementation when an interruption occurs, alerting management to the need for an emergency or standby power supply.

3.5.4 Alternates to Orderly Shutdown. An orderly shutdown may not be acceptable. Then an alternate may be selected such as:

(1) Maintain utilities without an interruption. Full capacity standby power must be available as well as uninterruptible power for boiler controls, on-line computers, and essential relays and motor starters.

(2) Accept an outage but return on standby power. Full capacity power must be available. Startup of functions will be required since magnetic motor starters and relays will have dropped out. Although power is off for 0.1–5 s, it will take 15 min to ½ h or more for return to normal operation.

3.6 Heating

3.6.1 Maintaining Steam Production. Continuous-process plants require uninterrupted steam production. Minimum requirements for continuous steam production are sufficient combustion air, air to instruments and actuators, water and fuel supplies, plus a continuous power supply to most flame supervision systems. The maximum interruption tolerable is that duration during which the inertia of the fans or pumping equipment will maintain flows and pressures above minimum limits. Table 4 illustrates how this can be achieved.

3.6.2 Process Heating. Process heating is defined as heat required to maintain certain process materials at the required temperature. Noncritical heating processes, due to the inherent heat capacities of such systems, can withstand a power interruption of considerable duration, say, 5 min to a maximum of several hours.

**Table 4
Systems for Continued Steam Production**

Components	Allowable Outage Duration	Systems
Flame supervision systems	Nil	Mechanical stored energy systems Motor-generator set ride-through UPS systems
Motor controls and instrumentation	Nil	Same as above
Boiler fans	½ to 2 s	Multiple utility services either on-line and relayed, or off-line, switched, and transferred
Air compressor	To 30 min, depending on storage; nonessential air users should be automatically shut off	Multiple utility services Turbine- or engine-driven generator, off-line Turbine (combustion, steam, water) off-line
Water pump	To 5 min, depending on water drum capacity and upsets in steam production caused by power disturbance	Multiple utility services either on-line and relayed, or off-line, switched, and transferred Turbine (combustion, steam, water) off-line, automatic start
Oil pumps (for burners)	To 15 cycles, more with flywheel	Multiple utility services, on-line and relayed Turbine (combustion, steam, water) on-line
Electric oil supply pumps	Several minutes	Turbine- or engine-driven generator, off-line Multiple utility services, on- or off-line

Any of the following systems would be adequate for this application:

(1) Engine-driven generators, off-line

(2) Multiple utility services on-line and relayed, or off-line, switches, and transfers

(3) Turbine (combustion, steam, water) off-line or on-line

(4) Mechanical stored-energy system, with auxiliary motor-generator sets off-line

Other heating processes such as cord and fabric treating and drying are of such a critical nature that loss of heat will cause an out-of-specification product within 10 s, but the gas or oil burners and flame detectors are sensitive to drops in voltage of about 40% for a second or less. In this case, an uninterruptible supply for the controls and an engine generator for a 10 s main power supply or an alternate feed system may be required.

Infrared drying of enamel on automobiles and appliances is a form of heating by electric energy that must be maintained. A short interruption, perhaps 10 s, may be acceptable to bring on-line standby generation or perform automatic switching to an auxiliary power source.

Losses may be substantial should power be lost during the heat treating of metals and when either direct or indirect melting of metals is in process. Two evaluations should be made: one based on not losing the flame on the fuels used, and the other based on accepting an interruption for a short period with a restart necessary. Uninterruptible power is several times more expensive than switching or emergency power, but saves the product and prevents process interruption. Switching or emergency power is less expensive, but product or process losses may exceed the initial savings in the chosen electric supply system.

Induction and dielectric heating are other forms of electric heating that may or may not allow short interruptions to be tolerated. Most processes, whether in production or in other fields, could tolerate an interruption of sufficient duration to come on-line with switched sources or an engine- or turbine-driven generator.

3.6.3 Building Heating. Buildings, even in the coldest regions, can usually be without heating for a minimum of 30 min. The same systems listed for noncritical heating processes will be suitable for this application.

3.7 Refrigeration

3.7.1 Requirements of Selected Refrigeration Applications. Requirements for refrigeration are usually noncritical for short power interruptions of several minutes to several hours. The need may become extremely critical as the length of time of the outage increases. Consider these refrigeration needs:

(1) Production of ice cream or the freezing of foods may stop in the middle of the process. Not only will all production be lost during a power failure, but damage may result to the product in process.

(2) Material in storage may be in jeopardy as temperatures rise. Cafeterias, frozen food lockers, meat cooling and storage facilities, dairies, and other food operations require refrigeration and will soon be in trouble as the length of the power outage increases.

(3) Scientific tests of long duration may require accurately maintained low temperatures. Short outages of electric power may destroy the tests and require repeating. Any expensive and time-consuming process should be provided with a standby power system.

(4) Medical facilities require refrigeration for blood banks, antibiotics, and for long-range laboratory experiments and cultures that could be spoiled.

(5) As the state of the art of superconductivity develops, improves, and spreads to applicable fields, power for cryogenic refrigeration equipment operation will probably become critical.

Where present refrigeration units are electrically driven, when new units of moderate size are to be installed in permanent locations, and when other needs exist (as is usually the case) for emergency or standby power, a common engine-driven generator or alternate utility source should be considered.

Because of the slow rise in temperature of cold-storage facilities, a savings may be made by the use of a smaller than normal standby electric generator. By switching power to various units in turn, an acceptable storage temperature may be maintained until normal electric power has been reestablished.

3.7.2 Refrigeration to Reduce Hazards. Certain chemical processes are exo-thermal and release heat during the chemical reaction. Loss of the cooling or refrigeration system may cause severe damage or even an explosion.

3.7.3 Typical System to Maintain Refrigeration. A manual starting of an engine-driven generator, turbine, or alternate utility supply will usually suffice, providing a suitable alarm is installed to notify responsible persons of a loss of refrigeration.

3.8 Production

3.8.1 Justification for Maintaining Production in an Industrial Facility. Loss prevention in production facilities due to a power failure is justified by the total sum of many tangible and intangible savings. Some of these items to consider are as follows.

Is there a guaranteed wage clause in the labor contract? If so, there will be a direct loss in wages paid for which no production is received. Where power requirements are low and the heavy power demanding machines are left shut down until normal power is returned, a small electric supply system can be justified to supply finishing areas, inspection areas, office areas, and other areas where most of the people work.

Who is waiting for the product? There are periods when the products are being routed to warehouses and machines are not running at capacity. In this case the cost of machine downtime is lower than it would be for a product that a customer will not receive or will receive late if production is at full capacity and an outage occurs.

What is the cost of the product spoiled in process? In the rubber industry mate-rial may become sealed in vulcanizers, extruders, or mixers at high temperatures and be difficult and expensive to remove. Steam or water pressure may drop to zero and prevent proper cures with losses due to poor or ruined products.

If all electric power is lost during certain processes in the making of sugar, glass, steel, pharmaceuticals, rubber, paper, chemicals, and some other materials, the product may have to be scrapped.

What is the cost of consequential damages? While the material ruined may be scrap, there may be as many problems and costs associated with its removal as with the loss itself. Some material must be dug out, or removed by hand, piece by piece, until lines are cleared or chambers are empty and clean so that an orderly startup can follow.

What is lost in reestablishing work efficiency? A ½ h electric power interruption disorganizes the workers. Experience indicates that, following the interruption, it will take workers at least two or more hours to settle down, go to work, and reach the production level at which they were operating just prior to the power failure. It may take days to reestablish normal procedures in scheduling of incoming and outgoing materials and in telegrams, letters, notices, and calls explaining delays and changing promises.

A less tangible item lost is good will. For example, in the film processing industry the customer may not consider the replacement of the exposed film with a new roll of unused film as adequate compensation for his "once in a lifetime" pictures that were spoiled due to a power failure.

Real and potential costs and losses should be calculated or estimated and added together to justify an emergency and standby power system for industrial and commercial facilities.

A reasonable estimate of the costs associated with each past power failure should be calculated and recorded in a journal with the date, duration, and conditions existing at the time. As time goes on this could be valuable factual backup information for budget requests.

3.8.2 Equations for Determining Cost of Power Interruptions. A rough estimate of the cost of a power failure from a cash flow viewpoint may be calculated as follows:

total cost of a power failure = $E + H + I$

where

E = cost of labor for employees affected, in dollars
H = scrap loss due to power failure, in dollars
I = cost of startup, in dollars

The value of E, H, and I may be calculated as follows:

E = AD $(1.5\ B + C)$
H = FG
I = $JK\ (B + C) + LG$

where

A = number of productive employees affected
B = base hourly rate of employees affected, in dollars
C = fringe and overhead hourly cost per employee affected, in dollars
D = duration of power interruption, in hours
F = units of scrap material due to power failure
G = cost per unit of scrap material due to power failure, in dollars
J = startup time, in hours
K = number of employees involved in startup
L = units of scrap material due to startup

After the cost of downtime has been calculated the savings in utilities should be subtracted to arrive at a total cost of downtime.

3.8.3 Commercial Buildings. For commercial establishments, a similar example may be assembled based on the length of the power interruption, labor cost, loss of profit on sales, loss due to theft, and startup costs.

3.8.4 Additional Losses Due to Power Interruptions. In addition to losses relating to cash flow are those more difficult to calculate but which should be included when available and applicable, such as

(1) Prorated depreciation of capital costs
(2) Depreciation in quality in process materials
(3) "Cost" of money invested in unused materials or machines

Other losses may occur under special or unusual conditions. In an industrial plant operating at 100% capacity any loss in production results in the loss of the profit of the item or service. The prorated cost of fixed and variable overhead becomes a loss. Customers may switch to competitors. Expenditures for standby power have additional justification under this condition.

3.8.5 Determining the Likelihood of Power Failures. Next, the likelihood of a power failure should be determined by studying the record of the plant or utility company electrical supply, or by transmitting the service requirements to the local utility and obtaining their recommendations. Examples of recorded power failures are shown in Table 5.

Rather than complete power failures as recorded in Table 5, Table 6 covers short-term dips.

A projection should be made, working with the utility company, as to whether or not the power reliability will improve or decline. Since the cost of a power failure, as defined in this standard, is paid by the user, it is important that he relate the reliability of power duration and quality to the need for justification.

Table 5
Example of Recorded Power Failures

Date	Time	Duration	Transmission Line
9 March	09:52	10 min	14
11 June	21:53	12 s	14
11 June	22:13	9 s	14
15 July	20:40	5.5 s	13 +22
17 July	19:13	1–2 min	14 (9 times)
	20:44		

Table 6
Example of Recorded Short-Term Dips

Date	Time	Line	Duration (cycles)	Line-to-Ground Dip (per unit) E_a	E_b	E_c	Voltage After
14 April	21:50	32	18	0.86	0.81	0.75	1.0
30 April	15:53	System	43	0.83	0.92	0.92	1.0
9 May	07:52	23,24	32	0.90	0.85	0.83	1.0
9 May	07:54	20	24	0.31	0.31	0.58	1.0
9 May	07:55	22	46	1.00	1.00	0.69	1.0
						0.40	
						0.50	
						0.06	
17 May	00:43	13,22	21	0.50	0.69	0.31	1.0

3.8.6 Factors that Increase the Likelihood of Power Failures. As a full designed load is reached or exceeded, the probability of a power failure increases. A similar probability exists as systems become more complex and as system equipment ages.

3.8.7 Power Reserves. Power reserves in the user's area should be investigated. Adequate reserve margins above peak load demands provide a guide to power reliability because the margin provides for some contingencies.

3.8.8 Examples of Standby Power Applications for Production. The following typical examples were taken at random from users who found, after evaluation, purchase, and installation, that a standby power system was justified:

(1) A drug company installed a special engine-generator backup system based on the need for constant temperatures in manufacturing processes. The value of processes saved during blackouts soon exceeded the cost of the equipment installed by more than 10 times.

(2) A photographic film processing company specified a 45 kW engine-generator when the film processing machines were ordered. Film must be moved along in each aspect of the process within 30 s or loss of quality will result.

(3) A 400 kW, 480 V, three-phase engine-generator was installed in a highly automated egg farm. Electric service continuity was mandatory for egg production and the welfare of the "machines" (chickens). During the first year, two electric service outages occurred and the automatic emergency generator reliability supplied power.

3.8.9 Types of Systems to Consider. An engine- or turbine-driven generator or an alternate independent utility source usually will fulfill the requirements for standby power. More than a standby system may be required for critical loads. Motor starters, contactors, and relays held closed by a coil and magnetic structure are especially sensitive to short-time power outages or voltage dips. Their dropout characteristics vary with respect to voltage level and the length of duration of the voltage dip. As a guide, a voltage dip to 70% or 60% of rated voltage for 0.5 s will deenergize many of the devices. The longer the time of the voltage dip, the more devices will deenergize. Depending on the application of these devices, an emergency or uninterruptible power supply system may be justifiable, especially where boiler controls, critical chemical processes, safety devices, and other critical systems are required to be maintained [18].

Devices much more likely to fail or malfunction due to voltage excursions are modern static switching devices, which have almost no time delay, and pressure, temperature, and flow transmitters and receivers of the electronic type. Automatic calculators for process control, data reduction, and logging are in this same category [19].

An uninterruptible power supply system for critical on-line computer or boiler control loads may be required since a 30-cycle outage may mean restarting the plant. In combination, the uninterruptible power supply system may be of short duration since the standby equipment should be designed to be on-line in from 10–60 s.

Factory clocks used for production control, or for the basis of incentive wage payments, should be arranged to maintain accurate time by the use of batteries for power during the time of any power interruption.

Support of production facilities will justify most or all of the user's needs detailed in this chapter. The evaluation, justification, and decision to purchase and install a standby, emergency, or uninterruptible power supply system, or a combination of these systems, should include the consideration of all the electric power requirements for all listed needs in the case of a power failure.

3.9 Space Conditioning

3.9.1 Definition. Space conditioning is a controlled environment, either to maintain standard ambient conditions or some artificial alteration of a standard environment in a building, room, or other enclosure.

3.9.2 Description. A controlled environment may include any of the following variables:

(1) temperature
(2) vapor content
(3) ventilation
(4) lighting
(5) sound
(6) odor
(7) gas
(8) dust
(9) organisms

3.9.3 Codes and Standards. Codes apply primarily where personnel safety is involved, but still very little reference is made specifically to power supply requirements. Most requirements in OSHA [25] are dedicated to ventilation of areas where hazardous gas, contaminants, and other similar elements and conditions can exist that would endanger personnel safety. Sections 31.5518, 1910.1006, and 1910.96 of OSHA [25] give some such requirements. Where *adequate* ventilation is referenced, conditions exist that may warrant backup or uninterruptible power supplies to provide this ventilation.

3.9.4 Application Considerations. Air-conditioning loads for personal comfort are not normally considered critical and are even shed under overload conditions in some cases. Where equipment is sensitive to temperature, however, such as where solid-state electronic components exist, air conditoning can be critical. Where continuous backup or standby power is not available, a self-contained emergency or standby generator would apply. An uninterruptible power supply specifically for this purpose is not normally necessary since loss of power would cause no instantaneous temperature change. Likewise, where moisture and humidity cause serious equipment and operating problems, heating might require some type of backup power supply.

Table 2 shows examples of tolerable outage time for various applications. Ideally, tests should be run by the designer or planner to determine the maximum transfer times tolerable. Amount of power needed, acceptable outage time, and economics will determine the applicable power supply. Often economics dictate that power for space conditioning be incidental to total power supply requirements, and the user should evaluate the ultimate consequences of loss of power. An adequate alarm system might be a consideration in this case.

Alarm and signal circuits often require as much or more reliability than equipment providing critical space conditioning. An uninterruptible power supply would be necessary where even a temporary loss of power would endanger personnel or cause severe equipment damage.

3.9.5 Examples of Space Conditioning Where Auxiliary Power May Be Justified. Typical situations and facilities in which a comprehensive study of the need for supplemental electric power is warranted are as follows:

(1) Commercial and laboratory horticultural botanical installations may require programmed cyclic control of temperature, humidity, and lights to develop the crop yield or desired experimental results. The loss of temperature or humidity control for 6–8 h can result in total loss of a crop. A greenhouse must have maintained temperature in order to produce.

(2) Tropical animal raising requires control of ventilation, temperature, humidity, and lighting. All are completely dependent on electric power. A loss of heat or cooling can result in death or illness to all animals being raised. Lighting and temperature changes from the established cycle can induce unwanted breeding periods in many exotic creatures. Egg production may be greatly curtailed by temperature changes or loss of light.

(3) Agricultural operations are often located in remote areas where long utility lines are susceptible to damage.

(4) Final operations and packaging of materials susceptible to contamination are conducted in *clean room* type environments. A power interruption will shut down the total operation, and contamination may result as people exit, unless the room is kept under positive pressure to prevent in-drafts from bringing in contaminants. Such contaminants will necessitate a complete recleaning of the room before it can be used again.

(5) In large processing plants, priorities may be established for essential loads. Nonessential loads may be dropped automatically by load-shedding systems in the event of a power failure, and other limited emergency sources must be relied upon. A reevaluation of these priorities should be undertaken. Critical temperature controls may have been placed in an air-conditioned space. These controls must function to achieve an orderly process shutdown, but can fail due to overheating caused by lack of ventilation or cooling.

(6) Windowless buildings or inside rooms may become unsafe for occupancy during an extended power interruption. Power supplies to keep these areas ventilated should be installed if evacuation of all personnel is not acceptable.

3.9.6 Typical Auxiliary Power Systems. Needs may require one or more types of emergency and standby power systems available. An engine-driven generator or combustion turbine should be considered for these applications. Switching to a separate alternate electric utility line would involve a lower capital cost, should such a line be readily available. In addition, a short-time uninterruptible power supply system may be required for critical applications to supply power during engine startup or switching.

3.10 Fire Protection

3.10.1 Codes, Rules, and Regulations. Various codes, standards, laws, rules, and regulations contain either advisory or mandatory statements related to emergency and standby power systems for fire protection. See California Administrative Code [16], Title 24, Part 3, Basic Electrical Regulations (Article E700, Emergency Systems); ANSI/NFPA 101-1985 [11]; IEEE Committee Report [20], pp 19 and 20; and Katz [B8]. There appears to be more written on the subjects of wiring system reliability and needs than on the source of electric power supply or the checking and maintenance of the complete installation. All three are vital.

Article 760 of the NEC [9] lists some requirements for protecting circuitry in alarm and signal systems. Article 230-94 of the NEC pertains more to the power supply by allowing a fire alarm circuit to be connected to the supply side of a service overcurrent device if it has separate overcurrent protection.

The requirements of all local, state, and national standards and codes should be determined. The insurance company that will underwrite the insurance can provide valuable assistance in making sure that all requirements are met.

A common-sense approach should be used even beyond meeting the letter of the law and insurance requirements as a minimum standard. The real goal is to avoid a destructive fire, or in the event a fire does start, that it be held to a local area with minimum damage to property and no harm to personnel. Knowledge in depth of the industrial facility and processes by plant engineers and other responsible plant personnel should be utilized to reduce the likelihood of a fire and the extent of damage should one begin.

3.10.2 Arson. Arson may be a source of a fire and the need for power for plant security, lighting, signaling, and communication covered in other sections should be regarded as contributing to the reduction of possible loss by fire from this cause.

3.10.3 Typical Needs. The user's possible specific needs connected with the general need of emergency and standby power systems for fire protection include

(1) Power, usually batteries, to crank the engine on an engine-driven fire pump

(2) Sprinkler-flow alarm systems

(3) Communication power to notify the fire department and to assist in guiding their activities

(4) Lights for the firemen to work by in the buildings, around the outside area, and mobile on company trucks

(5) Power for the boilers that supply steam-driven fire pumps

(6) Motors driving fire pumps, well pumps, and booster pumps

(7) Air compressors associated with fire water tanks

(8) Smoke and heat alarms

(9) Deluge valves

(10) Electrically operated plant gates, drawbridges, etc

(11) Communications, such as public address systems for directing evacuation of personnel

(12) Fire and hazardous gas detectors

3.10.4 Application Considerations. A fire almost always warrants initiation of an emergency shutdown in a plant either by operator-initiated devices or by automatic operation. The circuit required for shutdown is obviously critical as are all main circuits for the abovementioned fire protection equipment. Uninterruptible power supplies should be first choice for these applications. Where automatic fire protection is employed, like sprinkler systems, CO_2 discharge, etc, nuisance initiations should be prevented, a condition which lends itself to application of uninterruptible power supplies.

It is common practice in large plants to back up an electrically driven fire water pump with a mechanically driven one such as a diesel drive. If all ac power is lost in an emergency shutdown operation, fire protection is maintained. In this case, the ac power supply to the motor-driven pump is not absolutely critical. If a mechanical drive pump is not available, the motor-driven pump would normally be supplied by an emergency generator.

Under emergency conditions, such as fire, communications are sometimes vital to personnel and equipment safety. Under emergency conditions, communications should not be subjected to an outage of any kind. Even momentary outages of a few cycles might cause erroneous communications, especially where remote supervisory control is employed.

3.10.5 Feeder Routing to Fire Protection Equipment. Electric power distribution systems supplying fire equipment should be routed so as not to be burned out by a fire in the area they are protecting.

Protection of control circuitry, as well as main fire protection circuitry, can be enhanced by underground conduits, separate conduit and wire from other circuits, and fire-resistant insulated cable.

3.11 Data Processing

3.11.1 Classification of Systems. Most data processing installations can be grouped into two general classes of operation in accordance with their usage. These classifications are off-line and on-line. These categories will be helpful in identifying a data processing system's vulnerability to electric power disturbances, since an off-line process will rarely require power buffering or backup sources or equipment. Conversely, it is common for an on-line system to warrant the additional expense of buffering or backup equipment.

3.11.1.1 Off-Line Data Processing Systems. These systems are generally set up to perform one or more programs at a time in a sequential or batch mode. Usually such systems have a program operator for automatic processing of control cards for a given job run. Often, 24 h operation for a heavily loaded system may be necessary. In the off-line category, for the most part, are business, scientific, and computer center applications. Systems of this type are particularly vulnerable when the programs are lengthy (several hours in duration). Thus the insertion of several natural breaks or checkpoints in the program for segmentation of long programs is highly desirable. Programming can be designed to save intermediate results at a checkpoint and to have the option of restarting at the last checkpoint that preceded the power interruption. Such a practice in program interruption can be valuable in protecting against peripheral equipment failure. Many current programs are being designed without checkpoint techniques even though the practice has been found to be feasible. Data-dependent programs have running times that vary in duration with the magnitude of input data, and accordingly it is difficult to limit the run to much less than a 20 min period.

3.11.1.2 On-Line Data Systems. On-line data systems, or as they are often called, *real-time* systems, are systems that are time and event oriented. They must respond to events that occur randomly in time, often coincidentally. An awareness of the system of events that occur that are external to the computer and beyond its influence is a requirement. In this category are such applications as industrial process monitoring and control systems, airline passenger reservations systems, vehicular traffic control, certain specialized scheduling applications, international credit card and bank associated credit/transaction systems, plus many more. With these systems the computer outage problem due to power interruption is usually more critical than in off-line applications. Furthermore, there is generally no merit in segmenting programs. In most cases, any outage or power interruption will result in the loss of some data that were available only during the time period of the outage. The form of the input data is not conveniently available for a rerun, but may come from sensors such as thermocouples or pressure transducers that are scanned by a computer. When a computer controls a process, potential problems resulting from a power disturbance are generally serious enough in terms of product damage or equipment malfunctioning to warrant the use of a reserve or backup power source. Furthermore, any solution, to be adequate, must accomplish the necessary switching to the backup source without power interruption to the computer. It is obvious that the potential losses to several hundred users, or input stations, to a time-shared computer system would warrant the providing of a backup source that can practically guarantee uninterruptible power. In cases where equipment or process monitoring must be made, as through data logging, protection from destruction of only the core memory content during a power interrruption may be adequate. Automatic restart upon return of power is possible to minimize time that the equipment is down and can often be utilized with the additional provision for a manual means of updating the system data or information not gathered or scanned during an interruption.

3.11.1.3 Single-Phase Versus Three-Phase. The foregoing discussion has categorized data processing equipment and resulting electrical loads by their respective functional usage, that is, whether they are off-line or on-line systems. In most cases, large systems are involved. A further differentiation that can be made is by the magnitude of load. As with most power utilization equipment, smaller power consuming devices can generally be supplied from a single-phase source. The differentiation between single-phase versus three-phase power consuming systems is often necessary since the methods of protecting against input power disturbances and outages can be quite different for each system. Some of the data processing systems that use single-phase power will employ microprocessors or minicomputers. Others may consist of multiple single-phase load units distributed in their connection to three-phase power so as to achieve a reasonable load balance when all units are operating. This may result in load unbalance when some of the units are turned off.

In general, computers and peripheral units that draw less than 1.5 kVA will often be single phase. Those that draw more than 10 kVA often require three-phase power. In most cases single-phase loads can be connected to three-phase sources provided load unbalance at maximum load is not excessive, generally taken as 25% or less.

3.11.2 Needs of Data Processing Equipment from a User's Viewpoint. In general, data are gathered in analog or digital form and converted to one or the other system. (Refer to information in the preceding paragraphs.) These data in electrical form are then processed through one or several programs by an electronic computer, minicomputer, or microprocessor. The results are traced or printed out, or a signal is generated and fed back to form a closed-loop system that provides control to match preset conditions. Both types of outputs are commonly available and used.

3.11.2.1 Real-Time Data Processing. Both industrial and commercial users apply real-time data processing systems with a computer on-line. Failure of the data processing system often causes loss of valuable data or interruption of a critical process, either of which may result in extensive financial losses and require hours, days, or even weeks to fully recover. Failure of computers providing control may also jeopardize the safety of personnel. These failures can occur as a result of an electric power failure. Thus a reliable high-quality primary power supply is frequently justified to minimize loss of money and to prevent injury or loss of life.

3.11.2.2 Process or Operational Controls. A short list of control operations that frequently include data processing equipment, minicomputers, or microprocessors will suffice to alert users to the types of hazards and losses that may occur as a result of a power failure. Applications are categorized for both industrial and commercial classifications.

(1) *Industrial Applications*: materials handling; regulating control and status monitoring of pipelines, compressor, and pumping stations at remote locations; mixing compounds; grinding, drilling, machining; twisting textile fibers; steel mill processing; refining (oil); automatic testing and gauging; fabric processing; power flow and load dispatching; safety and security monitoring; process control and data acquisition.

(2) *Commercial Applications*: controlling elevators; automated checkstand combined with inventory control; newspaper production machines; airline passenger reservations; computer-aided emergency vehicle dispatching; environmental, life safety, security monitoring and control for buildings; typesetting; accounting; traffic control — cars, rail, and airplane; stock exchange and broker transactions; corporation computers for engineering, scientific, and business matters; hospital diagnostic equipment and individual intensive care systems; banking, financial, and credit transactions.

3.11.2.3 History of Developing Needs. With the advent of the electronic computer as a part of data processing and process control, an increased emphasis has been placed on the need for emergency and standby power systems to assure a continuous flow of energy. Coupled with this need is a superimposed problem, namely, the suppression of most short-time switching interruptions, voltage surges, dips, and frequency excursions. Often the suppression of transient disturbances and the need for emergency or standby power can be satisfied through a single installation of supplementary or auxiliary equipment. These transient disturbances have been a part of the electric power supply in the past, but caused few problems until electronic equipment came into extensive use. In some instances the solid-state power control equipment has caused problems, particularly on small, independent power systems.

3.11.3 Power Requirements for Data Processing Equipment. By the proper selection of an electric supply system, the power needs associated with data processing with a computer can be met, namely, a reliable source of noise-free electric power at all times and of a much higher quality than previously demanded by most devices.

The problem is how to reconcile commercial short-duration power interruptions with relatively short time domains in electronic circuits. In early stages of data processing equipment and computer equipment development, it was not unusual to experience problems with hardware and software when power disturbances of microseconds in duration were experienced. Most equipment built in that era was extremely vulnerable to such short time disturbances. The goal of most manufacturers in today's technology is to build from 4 ms to 1 cycle of carryover, or ride-through, time into their equipment.

Table 7 shows computer input power quality parameters for several manufacturers. The user should consider Table 7 only as a source of some examples since computer designs vary with size of computers, their processing power, and the technology available when the design was created. They are continually changing and the parameters of power needs are changing rapidly with the designs. Some of

Table 7
Typical Range of Input Power Quality and Load Parameters
of Major Computer Manufacturers

Parameters*	Range or Maximum
1) Voltage regulation, steady state	+5, −10 to +10%, −15% (ANSI C84.1−1970 is +6, −13%
2) Voltage disturbances Momentary undervoltage	−25 to −30% for less than 0.5 s with −100% acceptable for 4 to 20 ms
Transient overvoltage	+150 to 200% for less than 0.2 ms
3) Voltage harmonic distortion †	3−5% (with linear load)
4) Noise	No standard
5) Frequency variation	60 Hz ± 0.5 Hz to ± 1 Hz
6) Frequency rate of change	1 Hz/s (slew rate)
7) 3ϕ, Phase voltage unbalance ‡	2.5 to 5%
8) 3ϕ Load unbalance §	5 to 20% maximum for any one phase
9) Power factor	0.8 to 0.9
10) Load demand	0.75 to 0.85 (of connected load)

*Parameters 1), 2), 5), and 6) depend on the power source, while parameters 3), 4), and 7) are the product of an interaction of source and load, and parameters 8), 9), and 10) depend on the computer load alone.

† Computed as the sum of all harmonic voltages added vectorially.

‡ Computed as follows:

$$\% \text{ phase voltage unbalance} = \frac{3(V_{\max} - V_{\min})}{V_a + V_b + V_c} \cdot 100$$

§ Computed as difference from average single-phase load.

the paragraphs that follow expand on the individual parameters that are presented in the table. Although there is a degree of variance among computer manufacturers, the following represents the principle power parameters that are considered important by most major companies. While several of these parameters, such as frequency variation, can be relatively insignificant when power is derived from a commercial power source that embodies vast tie networks, they can become an important design consideration when supplemental or independent power sources are applied as a means of power quality improvement.

3.11.3.1 Steady-State Voltage. 208Y/120 V single- and three-phase voltage is the most common computer unit utilization voltage with some single-phase 120, 120/240, or 240 V. Some equipment is reconnectable for use at several voltages by use of an internal tapped transformer. Tolerance on the 60 Hz voltage varies among manufacturers; however, limits as listed in ANSI C84.1-1982 [1] are + 6% and – 13%.

Systems requiring 400 Hz power (for mainframe and some peripheral devices) often derive their power from a motor-generator set that is usually an integral part of the computer system. The motor of these units, ranging from 10–20 hp, is ordinarily connected to a power source that is separated or isolated from the computer equipment. A common nominal line voltage is 480 V, three-phase, for which the rated motor-generator input voltage is $460 \pm 10\%$. The output voltage of the generator is closely regulated, usually to within $\pm 2\%$, and is distributed over special 400 Hz lines to the load. Smaller computers and peripherals may obtain power from within the computer unit or a nearby unit from small inverters within the equipment. These units are usually not greater than 20 kVA and are powered from the input 208Y/120 V three-phase service or a 120/240 V single-phase device. Since inverter characteristics during abnormal operations, such as inverter failure, are quite different from motor-generator set failure characteristics, it would be well to check with the computer equipment manufacturer before investing in supply equipment.

A growing trend is to provide regulated low-voltage direct current for logic circuits by rectifying the output from 400–40 000 Hz static inverters. These devices, though lacking ride-through ability when installed without the addition of sizable banks of energy storage capacitors, do offer a compact and convenient source of regulated low-voltage dc power. With proper design and redundancy, their reliability is at least equal to that of a central motor-generator.

distributed to computer equipment from a location remote from the computer equipment, special considerations must be made for voltage drop (increases with higher frequency systems). Conductors are ordinarily installed in nonferrous raceways.

3.11.3.2 Voltage Transients. Computer manufacturers usually specify maximum momentary voltage deviations within which their equipment can operate without sustaining errors or equipment damage. The transient conditions are defined in terms of amplitude and time duration. Historically, an example was a range from $\pm 5\%$ of undefined duration to –30% for 500 ms, and +130% for 5 ms. Another example is $\pm 20\%$ for 30 ms. A few manufacturers also specify a duration limit for total voltage loss of from 1 ms to 1 cycle. A figure of 8.3 ms (half-cycle) for older equipment was typical. These values should not be confused with impulse tolerances, which are of much shorter duration (microseconds) and higher level (500%) and are usually part of the noise susceptibility and electromagnetic compatibility (EMC) tests.

Figure 4 shows an envelope of voltage tolerances that is representative of the present design goal of a cross section of the electronic equipment manufacturing industry. Shorter duration overvoltages have higher voltage limits. Some computer manufacturers specify a maximum allowable limit for volt-seconds, typically 130% of nominal volt-seconds (area under the sine wave).

3.11.3.3 Frequency. The manufacturer's tolerance on 60 Hz equipment ranges from ± 0.5 Hz to ± 1%, with the majority of the equipment limited to ± 0.5 Hz. Time-related peripheral devices are most sensitive to frequency (clocks, card readers, magnetic tapes, disks). Ferroresonant (ac/dc) supplies, which are widely used by some manufacturers in peripherals, are also frequency sensitive since they operate

**Fig 4
Typical Design Goals of
Power-Conscious Computer Manufacturers**

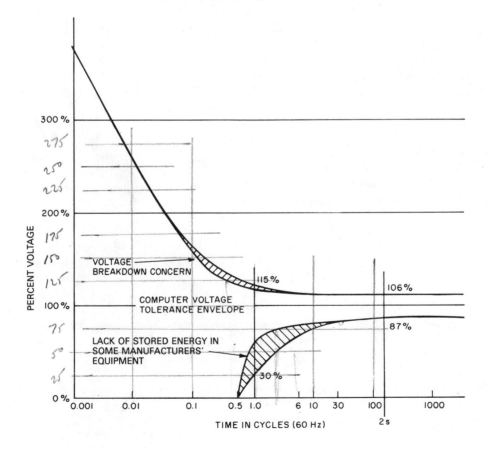

on a tuned circuit principle. They can generally tolerate variations of only ±1%. Some auxiliary motor-generators and other power supplies can tolerate variations as wide as ±3 Hz; however, for satisfactory operation of the entire system, the ±0.5 Hz tolerance should be maintained at all times for all parts of the system. Deviation from the tolerance may cause equipment malfunction or damage. 400 or 415 Hz power is used primarily in computer mainframes or in other areas where high-density power is needed. The higher frequency allows design of a smaller and more compact power supply and components that result in reduced heating losses. The deriving of 400 Hz power does involve additional costs. Also, distribution of 400 Hz power presents some special problems. Thus, for many of the peripheral manufacturers, 60 Hz power is still preferred. Some equipment, such as large CPUs, use both 400 and 60 Hz power, with 60 Hz mainly for cooling fans and blowers. The current trend is to use 400 Hz power in the larger data processing systems, with 60 Hz limited to the smaller systems and peripherals. Development and wider usage of 60–400 Hz or higher frequency static inverters to replace motor-generator sets and to be placed in individual equipment would reverse the trend back toward 60 Hz input in equipment. Several major computer hardware manufacturers predict the use of frequencies considerably above 400 Hz for future generation computer systems. Inverter frequencies of 20 kHz are commonly used now and some state-of-the-art inverters operate at 100 kHz and higher.

The voltage quality and characteristics of 400 Hz power are extremely critical, with frequency generally not as critical a parameter. Most manufacturers who consider the 400 Hz source an integral part of the computer would prefer to furnish conversion equipment as part of the computer installation. Furthermore, because of the special problems associated with the distribution of 400 Hz power, conversion equipment is generally located relatively close to the utilization equipment.

In addition to being sensitive to the limits of ±0.5 Hz, some system peripherals are also sensitive to the rate of change within this band. Although extensive information is not available, a variation of 0.05 Hz/s has been cited by one manufacturer as a limit for some units of the system. A typical limit is 1.5 Hz/s, measured as rate of change in a 10 cycle running average. The limit is most significant when turbine or engine-generators are applied where small load step applications may cause the rate to be exceeded.

3.11.3.4 Distortion. The maximum harmonic distortion permitted on input lines ranges from 3–5%, with the majority at +5%. The percentage is usually specified as total line-to-line distortion, with a maximum of 3% for any one harmonic.

Excessive harmonic content can cause heating in magnetic (iron) devices, such as transformers, motors, and chokes. The harmonic distortion will also appear as additional ripple in the output of some ac/dc power supplies and also cause threshold limits to vary in peak and average sensing circuits. Either can contribute to data errors.

It should be noted that elements of the load may introduce considerable distortion or *noise* into the power source. This reflected *noise*, although not in the source, may require suppression through filtering to avoid interference with other loads also connected to the system.

The voltage envelope should be sinusoidal, with a crest factor of 1.414 ±0.1. Waveform deviation should be limited to ± 10% line-to-neutral. The variation in the amplitude (in time) of the wave should not exceed ±0.5%.

Excessive modulation of the voltage can produce pulsing and speed variations in motors and can introduce additional ripple in the output of ac/dc power supplies.

Some computer units have half-wave rectifier units and SCRs (half-wave phase control). These are capable of creating a dc component of load current and greater current in the neutral than in the phase conductors. Power sources for such loads must be capable of handling them.

3.11.3.5 Phase Angle Displacement. Although some manufacturers build only single-phase equipment, the majority of equipment, particularly for larger systems, is three-phase. The maximum deviation from normal 120 ° spacing ranges from ± 2.5° to ± 6°. Unequal phase displacement, whether in the source or due to unequal loading, can further contribute to phase voltage unbalance.

3.11.3.6 Voltage Balance. Although not specified by all manufacturers, the maximum phase-to-phase voltage unbalance (defined below), with a balanced three-phase load, should be in the range of 3%. Unequal distribution of single phase, as is frequently encountered in computer systems, will increase the amount of voltage unbalance. Wherever possible, effort should be made to distribute load evenly between the phases.

Excessive phase voltage unbalance can cause excessive heating to three-phase devices such as motors. Similarly, relays and other electromechanical devices may be damaged due to continuous operation at high (or low) voltage. In addition, high *ripple* may be observed in some three-phase ac/dc power supplies if the voltage unbalance to the supply is high. Percent voltage unbalance is defined as

$$\frac{3(V_{max} - V_{min})}{V_a + V_b + V_c} \cdot 100$$

3.11.3.7 Environmental Considerations. It is well known that electronic equipment is capable of reaching design performance levels only when proper cooling is introduced. The principal thrust of this discussion involves the electric power needs of data processing equipment. However, it should be recognized that cooling of that equipment can be an overriding and sometimes limiting factor in the operation of data processing equipment, principally because most cooling and ventilating equipment is electrically powered. In addition to comfort air conditioning for the computer room, the computer's central processing unit (CPU) logic and chassis often have special cooling requirements, and many have overtemperature alarms or cutoffs. With larger systems and increased capabilities, computers have, for technical reasons, also grown more compact, leading to increased power densities in terms of watts per cubic foot. The result has been a progression in chassis cooling from natural convection, to forced air, to chilled water, and in some cases, to a refrigerant-cooled chassis. One manufacturer uses a dual freon unit for cooling the mainframe alone. However, intermediate-sized computer installations are frequently cooled by forced air through ducts or under floor areas. Small computers and peripherals may contain their own internal fans and draw cooling air from the room where they are installed.

The methods of cooling of a computer system can also heavily influence how long a computer system can operate without electric power to cooling apparatus. Assume that computer power could be maintained through the use of batteries, standby generators, or other auxiliary equipment, which is described later. It then becomes important to consider the design/selection of computer system cooling systems and their need for auxiliary power supplies. Obviously, there is no need to provide battery ampere-hour capacity, which can extend computer operating time beyond the time that a computer system can operate before it must shut down due to overheating.

Most manufacturers are reluctant to cite these times for their computer hardware since many variables within a facility can exist. For forced-air cooling, many claim up to 15 min can elapse before overheating and equipment shutdown occurs due to operation of overtemperature sensors. For those systems that use chilled-water cooling, approximately 2–3 min has been given as the length of time that the chilled supply can be cut off. For refrigerant cooling, about 15 min is often considered the maximum.

Most larger systems shut down automatically when a high temperature is reached and many provide a warning alarm when approaching a high temperature.

3.11.3.8 Humidification. The need for, and extent of, introduction of humidification into a data processing equipment room will vary considerably with the geographic area. Humidity control is required to ensure orderly movement of papers, cards, and magnetic tapes in the operation of the computer and its peripherals. Low humidity allows the building up of static electrical charges, which in turn causes cards and tape to stick together, jam, etc. Extremely high humidity can result in condensation of moisture on chilled chassis plates, resulting in rust and corrosion. Most data processing equipment manufacturers recommend that a humidity range of 40–60% be maintained in equipment rooms. As with cooling apparatus, discussed in the preceding paragraph, humidification generating equipment can be as important as cooling equipment. Any reserve or standby equipment should be sized not only for the supply of electric power to data processing equipment, but also for the length of time the equipment can operate without supplemental humidification or dehumidification and attendant problems.

3.11.4 Factors Influencing Data Processing Systems on Incoming or Supplementary Independent Power Sources. Much of the foregoing discussion has been dedicated to the various influences that the source of electric power has on data processing equipment and systems. It should be recognized that the same equipment and systems can, in various ways, interact with and influence the source power system and other utilization equipment served by it. Some of these influences become important if supplementary independent power sources are used as an enhancement of power source quality, particularly where emergency and standby power is applied.

3.11.4.1 Load Mangitude. The electrical load of a data processing system depends mostly on the makeup, complexity, and even function of the system. A typical small system may vary from 10–50 kVA, and larger systems from

100–300 kVA. Some multiple systems may be as large as 2000 kVA. As mentioned in the foregoing paragraphs, as systems become larger, some manufacturers elect to power mainframes with 400 Hz power derived from a 60 Hz source, often through separate motor-generator sets.

A principal computer manufacturer states that approximately 30–35 W per square foot per computer floor space should be allowed in planning for ultimate computer electric power consumption. This load density is exclusive of any attendant space conditioning, humidification, or dehumidification loads that would be required.

3.11.4.2 Load Growth. For an established system, most load growth will result from the incremental and often unpredictable addition or expansion of peripheral devices (tapes, disks, cardpunch, etc). However, not all system growth results in electrical load growth. For example, the replacement of several smaller disks by a larger (in a computing sense) disk device may actually require less power. Major load growth occurs when an entire system performance capability is upgraded, which usually results in a new mainframe and some peripherals.

3.11.4.3 Power Factor. The characteristic power factor of a computer system is relatively high. The power factor of the 60 Hz portion of the load generally ranges from 80–85%, and for the 400 Hz (motor-generator set) load generally ranges around 90%. Depending on the amount of 60 Hz load, and if a motor-generator set is used, the overall combined power factor is usually in the range of 80–90%. During initial powering up or startup, the power factor may drop as low as 50% for short periods.

3.11.4.4 Load Unbalance. Most equipment manufacturers attempt to balance their equipment load in the design of the equipment and in load connection in the planning phase. However, load unbalance may run from 5–30% (phase-to-phase) steady state and up to 100% dynamic, as in startup. The effect of load unbalance is to produce unbalanced phase voltages. See the discussion of voltage balance in the preceding paragraphs.

3.11.4.5 Startup. The powering up of a computer system may place severe demands on the power source. In that regard independent power sources are more vulnerable than a commercial power source from the utility company. Efforts are made by manufacturers to reduce inrush by various methods. Energizing of large loads is often sequenced in steps, manually or automatically. Special motor starting techniques can be used to reduce inrush in the starting of large motors, disks, and motor-generator sets.

The user often can employ operational procedures designed to ease startup and step loading. As an example, large groups of peripherals may be started manually, in sequence, rather than simultaneously. Large 400 Hz loads, such as mainframes powered by motor-generator sets, may be brought on-line slowly by controlled buildup of the generator output to give the logic load a cushioned start as well as reduced inrush.

Even with reduction methods, high inrush currents are common with many pieces of computer equipment. One manufacturer's central processor whose steady-state load is 24 kVA presents a 1500% (of steady state) transient for approximately 100 ms, which decreases to 600% in 300 ms. Another manufacturer, requiring 60 kVA steady state, presents a 400% transient. Still another manufacturer requiring about 8 kVA steady state presents a 1000% transient for a half-cycle, dropping to normal not later than the next cycle. The startup of a 40 kVA motor-generator set can require up to 200% for 2–10 s, even with special starting methods. High inrush puts an added requirement on the design of the power source, especially on devices such as ride-through motor-generator sets and static inverters. Current-limiting protection for inverters is typically 125–175%. Voltage output will drop during current-limiting output.

3.11.4.6 Step or Pulsing Loads. Even when a system is on-line and drawing steady-state power, it may present severe load transients due to the operational demands of the computer. Step changes in load as high as 200–300% may be possible when starting an additional unit. Frequently, step changes in load can be minimized within the equipment through judicious programming that avoids simultaneous energizing or operation of several processes or pieces of equipment.

Pulsing loads can cause problems and occur when a number of devices that have power peaks, which are coincident and repetitive, are connected to the same power source. As an example, the programming of multiple tape units to rewind simultaneously should be avoided. Line printers whose steady-state load is small may step 100–200% when striking a full line of print. The cumulative effect of a large number of printers in synchronism can place a severe strain on the power system. If the power source includes a rotating device with speed control, it is possible for the load to cause oscillations in the control and output. Although extremely rare, the problem should be considered if there could be a large number of synchronized pulsing loads.

3.11.4.7 Load-Generated Harmonics and Noise. Certain elements in the load, such as saturated magnetic circuits (transformers, motors), may cause distortion of the voltage waveform. Problems often result from the reflection into the source of load-generated noise, such as switching spikes caused by turning devices on or off or by the firing of high-speed solid-state devices (SCRs, diodes) that are a part of the computer load. The spikes, of microsecond width, may run several hundred volts on the 120 V line. These disturbances may have to be eliminated by filtering to avoid interference with other parts of the load.

The load factor for an on-line system may approach 100% due to the continuous nature of its operation. Off-line systems will generally be lower, depending on their daily scheduling and use.

Demand factor is the ratio of the actual steady-state running load to which a power system will be subjected to the total connected load.

Demand factor is important since connected kVA loads are generally available from the equipment manufacturer as nameplate ratings of individual components and can be taken as the worst-case condition. Thus, an arithmetic total of the loads can be computed. However, some computer system manufacturers provide data on the actual running loads, as seen by the total system, when operating the individual components as a system.

Through experience and data obtained from several of the computer systems manufacturers, some of the larger computer systems having connected loads in the area of 300 kVA experience demand factors of between 75% and 85%. In other words, the running loads can be predicted to run between 225 kVA and 255 kVA for such a system. Thus users should be aware that nameplate ratings indicate power for a fully featured machine and actual loadings may be significantly less for individual user's systems.

Larger systems, consisting of many components, will probably have lower demand factors when compared with smaller systems with fewer components because of inherent diversity among components.

3.11.4.8 Grounding. Most computer manufacturers have preferred methods of system and equipment grounding for their hardware. Some systems require special grounded signal reference grids with single-point grounding and strict control of the paths through which equipment ground and power system ground are interconnected. These grounds are always connected one to the other to be electrically safe and to conform with the codes. The paths may be separated up to some point. Also, since computers and office machines are accessible to people, particular attention should be given to the safety grounding of all equipment. If, for the sake of radio frequency noise reduction, the equipment grounding green wire is isolated from the building/raceways, it must be run with power conductors and carried back to the service equipment grounding conductor. Furthermore, special semiconductive flooring may be employed for computer rooms, which enables draining off of static charges and minimizes the shock hazard to personnel. Some equipment manufacturers recommend grounding of equipment to the floor grid. The manufacturer's installation instructions are not always correct and may contribute to problems. See Chapter 7 for grounding of emergency and standby power systems.

3.11.5 Justification of Supplemental Power. Insurance companies provide loss of power and loss of production insurance that can give a cost guideline on how much can be spent to improve the power supply.

The various units of a computer system are not equally sensitive to power supply disturbances. Power requirements for such equipment as air conditioning, lights, and drive motors are much less restrictive than for the logic, memory, control disk, and tape units. For economic reasons the power loads should be separated into those requiring buffer or filter action with an uninterruptible power supply and those that require no buffering and can accept a 0.5 s – 1 min power interruption while switching to a standby source of electric power.

Assuming that such a separation of computer or data processing equipment, or both, is possible, the next step would involve the selection of a system or equipment that would best solve the user's problems.

In summary, the following steps are intended to serve as a checklist toward problem solving where power problems are present. Care is required and the computer manufacturer's advice should be heeded in arranging and coordinating the ground references where more than one power source supplies power to computer units in a common system.

(1) Survey power quality requirements of all installed equipment to determine maximum degree of sensitivity of the individual components. These data are often not available from manufacturer's published literature.

(2) Check to see if the local utility company has operating records. If no records are available, the user may choose to install a power disturbance monitor to detect

and record transient power deficiencies at the point of usage and obtain a profile of power transients for a minimum two-week period. The results of such recordings should be used with the recognition that such seasonal conditions of lightning and other weather-related problems may not have been accounted for.

(3) Review and analyze the monitor's records to identify the magnitude, number, time, date, and characteristics of the recorded transients, interruptions, and prolonged outages.

(4) Classify the transients according to their origin: from the utility; from the operation of electrical equipment in the computer's vicinity; from the operation of computer units and their accessories.

(5) As appropriate, prepare engineering design and cost estimates of a power-buffering system of required capacity (kVA). The chosen power-buffering system should accommodate both future expansion and possible replacement with more recently marketed computer models.

(6) Evaluate actual and intangible losses due to interruptions or malfunctions of the electronic data processing equipment from all causes, including losses attributable to power supply deficiencies or failure.

(7) Compare expected losses without power-buffering equipment to the cost of owning and operating various types of power-buffering systems as insurance against transient-caused interruptions or malfunctions of computer equipment.

(8) If the previous step justifies an investment in a power-buffering system, proceed with the project.

3.11.6 Power Quality Improvement Techniques. Any attempts to categorize power improvement equipment should logically attempt also to define the time duration effectiveness of the equipment. The simplest and least costly equipment is limited in effectiveness to extremely short duration power disturbances. As the time span for protection of disturbances increases to *outage* type disruptions, the sophistication and associated cost for related equipment, as a general rule, will also increase. The following broadly classifies equipment from the very short duration to an indefinite time period. A broader classification of equipment could be the distinction between static and dynamic systems.

Equipment Systems:

Power conditioners
Short time (up to 15 s, but typically 0.1 s) ride-through
Extended time (up to 30 min) ride-through
Indefinite time

The time interval for most of the equipment listed can usually be extended to an indefinite time period through the use of a supplemental standby power source such as a second utility company power feeder or on-site generation. To maintain power input to the load within tolerance, the supplementary equipment must be capable of *riding through* the inherent sensing and transfer time switching associated with a second incoming feeder. With standby generation systems, ride-through time must at least equal prime mover starting time, including synchronizing and transfer switching times for those systems.

Short-time ride-through can be accomplished with *mechanical stored energy systems*, which are described in 5.2.

Extended ride-through time can be achieved through operation of the equipment in conjunction with stored energy equipment, such as batteries.

In general, *power conditioning* equipment is limited to improvement of short-duration power problems. In installations where electrical *noise* and voltage are principal problems, an isolation transformer can be an effective solution in noise attenuation and voltage control.

Isolation transformers can simultaneously isolate and change voltage, for example, from 480 V three-phase to 208Y/120 V, or can have a one-to-one transformation ratio. An ordinary transformer with separate primary and secondary circuits will provide some isolation. However, effectiveness is greatly improved if the circuits are equipped with special shielding between the primary and secondary windings. This special shielding will reduce the noise amplitude and inhibit the passage of noise through the transformer. Performance of three-phase transformers will be greatly enhanced in handling harmonic currents and unbalanced load if the transformer is connected delta-primary/wye-secondary on a three-legged core. Such transformers can be equipped with primary taps to adjust output voltage when input is constantly high or low. A typical tap range is +5% to -5% in 2.5% steps.

Beyond the isolation transformer is a wide range of voltage correction or stabilizing equipment, each having distinct and unique operating characteristics. Commercially available equipment that would fall into the power conditioner classification is listed in Table 8. Typical reported response times are listed for each device along with the approximate uninstalled cost or range in 1978 dollars. It is obvious, because of the relatively slow response times, that the use of several of the devices would be restricted to simple voltage regulation and would have questionable value against short-term disturbances. In addition, some of the devices have inherently high internal impedance. Any sensed voltage drop is ultimately corrected by its voltage regulating capability. However, under stepped load change, the dynamic voltage change may be unacceptable for the time required for the regulator

Table 8
Performance of Power Conditioning Equipment

Equipment Type	Response Time	Typical Internal Impedance (%)
Regulator, low impedance	11 ms	3–5
Transformer, constant voltage	25 ms	
Reactor, electronic saturable	30 ms	20–30
Regulator, electronic magnetic	160 ms	
Regulator, electromechanical	5+ s	3–5
Regulator, induction	5+ s	10
Tap changer, mechanical	60 s	3–5

to respond. If multiple loads are to be supplied by one device and one of the loads is switched, one must be assured that the magnitude of voltage change before the regulator can react will be within acceptable limits.

Dynamic regulation and response time are the major considerations. It is stressed that because of relatively slow response times some of the equipment is effective only for voltage adjustment or correction. Also, steady-state regulation is normally published, but dynamic regulation to step changes in voltage input and load current is often omitted. Without a detailed analysis of equipment operating characteristics, often from actual test data, a user could mistakenly think that he is getting protection against short-duration line input voltage disturbances and the effects of changing load. In general, equipment is available to accept line voltage variations of a 15% overvoltage to a 20% undervoltage condition with output regulated to ± 5% of rating. Such a regulated output is generally sufficient for electronic data processing equipment. In a situation where there appears to be a need for an ac line regulator, the application should be reviewed with the manufacturer of the data processing equipment.

3.11.7 Selection Factors for Supplemental Power. The true nature of the data processing operation should be examined in combination with its extent of usage. On-line systems generally require a higher degree of power protection than off-line systems. Certainly, a 24 h around-the-clock operation would be more vulnerable than a one-shift-a-day operation.

Whether on-line or off-line, the consequences of a power failure should be determined in terms of cost, value of lost data, necessity of reruns, inconvenience to and loss of revenue from customers, possible equipment damage, repair cost, and general annoyance. It is sometimes found, particularly in off-line operations, that the consequences of an outage do not justify the often considerable expenditure for a protective system. The decision should be based on how much a user should pay for insurance in light of the benefits to be derived.

To evaluate the effects of outages, an estimate of the number of outages and their duration predicted over a period of at least one year should be obtained from the utility, based on their past operating experiences.

A comparison of what is predicted with what is acceptable in terms of power limits should indicate if supplemental protection is needed.

3.11.7.1 Protection Time. Assuming that the predicted available power is not acceptable in terms of duration or frequency of disturbances or disruption, it should then be determined how much protection is appropriate and what solutions are available to the user.

Table 9 has been prepared to indicate what effect three types of power line disturbances will have on data processing and computer equipment. Also shown for each type of disturbance are several of the available solutions that can be applied for enhancement of the power quality. Note that in this comparison the disturbance durations are for less than 1 cycle, from 1 cycle–10 s, and for more than 10 s. Table 10 presents similar data in a slightly different manner and shows, for each of the categories of disturbances, how effective each of the available solutions will be as a power quality improvement measure.

Table 9
Summary of Typical Power-Line Disturbances

Type of Voltage Disturbance	Voltage Level of Disturbance	Duration of Disturbance	Typical Effects on Computer Equipment	Typical Power Enhancement Projects
Outage	Below 85% V_{rms}	More than 10 s	Built-in voltage sensors will power down computer equipment in an uncontrolled manner. Processing is interrupted usually resulting in excessive restart/rerun time, possible loss of data, or damage to hardware.	Uninterruptible power supply system, standby diesel generators, dual power feeders, general improvements to power distribution system.
Momentary under- and overvoltage (sags and surges)	Below 85% V_{rms} and above 105% V_{rms}	From 16.7 ms (1 cycle) to 10 s	Equipment may power down depending on duration and magnitude of disturbance. If so, processing is interrupted usually resulting in excessive restart/rerun time. In severe cases loss of data and damage to hardware may occur.	Solid-state switching between dual feeders, motor generator sets, fast response line voltage regulators, balance computer load on three-phase power, improve computer equipment grounding, general improvement to power distribution system.
Transient overvoltages (impulses or spikes)	100% V_{rms} or higher (measured as instantaneous voltage above or below the line V_{rms})	Less than 16.7 ms (1 cycle)	Data disruptions leading to errors, unready indications, etc, may cause individual equipment to stop processing. However, direct effects on the system are not normally detectable. Rarely, a severe transient will cause equipment to power down. Damage to electronic components may also occur if the equipment is not properly grounded or otherwise protected from transient overvoltages.	Isolation transformers, transient suppressors, power-line filters, primary and secondary lightning arrestors, balance computer load, improve computer equipment grounding.

Table 10
Relative Effectiveness of Power Enhancement Projects in Eliminating or Moderating Power Disturbances (US Navy)

| Disturbance Type | Uninterruptible Power Supply (UPS) System and Standby Diesel Generator | Uninterruptible Power Supply (UPS) System | Dual Power Feeders | | Motor-Generator | Solid-State Line Voltage Regulator | Specialty Shielded Isolating Transformer | Suppressors, Filters, and Lightning Arresters | Balance Computer Load on 3-Phase Supply, Improve Grounding |
			Secondary Spot Network	Secondary Selective*					
Transient and oscillatory overvoltage	All source-caused transients and no load-caused transients	All source transients and no load transients	None	None	All source transients and no load transients	Most source transients and no load transients	Most source transients and no load transients	Most	Some**
Momentary undervoltage or overvoltage	All	All	None	Most	Most	Some (depends on response time)	None	None	Some**
Outage	All	Only outages of a duration equal to the discharge time of the battery	Most	Most	Only "brownout"	Only "brownout"	None	None	None

*Includes special application of a solid-state static switch between two independent sources.
**These improvements do not eliminate or moderate power-line disturbances, but they do make the computer equipment significantly less susceptible to under- and overvoltages. Assistance of the computer manufacturer is generally required to identify grounding problems.

3.11.7.2 Reliability. An indication of reliability can be expressed in hours of mean time between failure (MTBF). Data for calculating the MTBF can be obtained from ANSI/IEEE Std 500-1984 [6]. Equipment and reliability can be enhanced by redundant systems and components, and provision for isolation or bypass of defective parts or subassemblies.

A typical reliability prediction assumes that if the uninterruptible power supply system fails, there is to be a high probability that the static switch will operate and that main power of suitable quality will be available to handle the load. The probability of simultaneous failures of the uninterruptible power supply and static switch or main power yields an exceptionally low failure rate corresponding to MTBFs of 10 years or longer. However, the probability of switch failure is greater when it is called upon to operate than when quiescent. Also, the duration of the failure in a static switch may be extended due to an inability to detect failure.

As with reliability, data on equipment life expectancy are difficult to obtain. Often, because of the rapid advancement of data processing technology, the protected equipment obsolescence will generally occur before the life of the protecting equipment is exceeded.

Dynamic or rotating apparatus is capable of withstanding short time and sustained overloads better than static equipment. Furthermore, most static assemblies are more vulnerable to shock or high-inrush loads than is dynamic equipment. Dynamic or rotating equipment does not lend itself well to expansion through paralleling due to complex control problems. Conversely, most static systems are easily expanded through the addition of parallel modules or additional batteries.

Because some systems have inherent lower overall efficiencies, increased heat losses will be evident. The first cost of supplemental cooling and ventilating equipment, with ongoing operating and maintenance costs, should be considered. Furthermore, the need of additional auxiliary or standby power capability to drive the cooling and ventilating equipment, when normal power curtailment occurs, should also be examined.

3.11.7.3 Other Factors. Most equipment will occupy building interior space, which not only costs money but will require environmental treatment or tempering. Such space is often treated as leasable space, and as such its cost considered as an ongoing cost in any comparative analysis.

Delivery, setup, checkout, and debug time can often be an important and deciding factor when the dependence and prime emphasis is placed on having a data processing system or computer and its capabilities in service on an around-the-clock basis.

Some of the solutions for improvement of power quality will require the installation of an auxiliary power source with some form of prime mover. Other sections in this standard provide detailed guidance toward a prime mover selection. The most popular choice is the diesel engine, with gas turbine usage increasing.

Often it is to the advantage of the user to provide for a separation of the source for the data processing equipment or its power improvement equipment from the source for the facility's environmental equipment. Since the only objective of the auxiliary power source is to provide an almost immediate source of power, reliability in starting is the most important characteristic. Usually, the unit is called on to start after a sustained period of idleness. The time period for starting could be important since the equipment could be called upon to operate in conjunction with a stored energy type of ride-through system. Often data processing equipment is installed in a relatively quiet atmosphere so that whichever choice of prime mover is made, some form of acoustic treatment will probably be required.

3.12 Life Safety and Life Support Systems

3.12.1 Introduction. Emergency power for life safety systems is required in many different types of commercial and industrial facilities. For example, an alternative power source is needed to illuminate exits and to operate alarm systems in most public buildings. These needs generally relate to fire safety as set forth in ANSI/NFPA 101-1985 [11].

In contrast, emergency power requirements for life support systems, such as a heart-lung machine or medical diagnostic equipment, are generally limited to health care facilities. These requirements result from operational needs unique to the health care environment.

Emergency power for both life safety and life support equipment is addressed in this section. In doing so, the emphasis is intended to be primarily from the viewpoint of the user.

3.12.2 Health Care Facilities. Typical examples of both life safety and life support emergency power requirements are found in large health care facilities. An appreciation for both of these needs can therefore be obtained by reviewing a hospital's emergency power system.

Hospitals are becoming increasingly dependent on electrical apparatus for patient life support and treatment as illustrated in the following paragraphs.

Electronic monitoring systems are being used to conduct therapy for critically ill patients. In the operating suite during open-heart surgery, an electronic heart-lung machine maintains extracorporeal circulation. In the intensive care unit patients depend on ventilators powered by electricity. Cardiac-assist devices augment a patient's own circulation in the coronary care unit. Continuous lighting is needed to observe patients in primary care areas and power is needed to maintain refrigerated storage of vital supplies such as in blood and tissue banks.

In addition to these examples of direct dependency on electric power in patient care, the sustained loss of electric power can also result in an undesirable traumatic experience for some seriously ill patients.

Hospitals also must have a highly reliable supply of emergency power for life safety systems to ensure that the lives of sick or disabled persons are protected during emergencies.

Interruption of normal electrical services to hospitals may be caused by a variety of natural phenomena. These include storms, floods, fires, earthquakes, explosions, traffic accidents, electrical equipment failure, and human error. Several large area blackouts have been experienced in the recent past with prolonged local outages. As a consequence of the "energy crisis" the incidence of such failures is thought by many authorities to be on the rise. Alternative sources of power for supplying vital life safety and life support systems must be provided to protect patients relying on these systems.

3.12.2.1 Power Continuity Requirement. The acceptable duration for an interruption of normal power service to critical hospital loads is the subject of many state codes and regulations (see Table 1). National standards often referenced by the states and specifically addressing this issue are ANSI/NFPA 99-1984 [10] and the NEC [9], Article 517.

ANSI/NFPA 99-1984 [10] requires that all health care facilities maintain an alternate source of electric power. With few exceptions, this source must be an on-site generator capable of servicing both essential major electrical equipment and emergency systems.

For hospitals, ANSI/NFPA 99-1984 [10] provides the following criteria with respect to the emergency system: "Those functions of patient care, depending on lighting or appliances that are permitted to be connected to the Emergency System are divided into two mandatory branches: the Life Safety and the Critical. The branches of the Emergency System shall be installed and connected to the alternate power source . . . so that all functions specified . . . shall be automatically restored to operation within 10 seconds after interruption of the normal power."

To meet the "10 second criteria" the emergency system must include independent distribution circuits with automatic transfer to the alternate power source. Two-way bypass and isolation transfer switches are recommended for the emergency branches. Figure 5 shows the emergency system wiring arrangement from a typical hospital. The hospital emergency system installation must follow Articles 518 and 700 of the NEC [9].

The *life safety branch* of the emergency system, as described in ANSI/NFPA 99-1984 [10], includes illumination for means of egress and exit signs (ANSI/NFPA 101-1985 [11] requirement), fire alarms and systems, alarms for nonflammable medical gas systems (ANSI/NFPA 56F-1983 [8] requirement), hospital communication systems, and task illumination of selected receptacles at the emergency generator set location.

ANSI/NFPA 99-1984 [10] contains a complete listing of circuits to be connected to the critical branch feed areas and functions related to patient care. For most of these critical loads the "10 second criteria" is considered to be sufficient. However, an instantaneous restoration of minimal task lighting, using battery systems, is recommended in operating, delivery, and radiology rooms where the loss of lighting due to power failure might cause severe and immediate danger to a patient undergoing surgery or an invasive radiographic procedure.

Examples of the types of life support and life safety equipment available with built-in battery backup power are:

Life Support:

(1) Aortic balloon pumps
(2) ECG and EEG monitors
(3) Portable defibrillators
(4) Portable respirators
(5) Task lighting

Life Safety:

(1) Fire monitoring and alarm systems
(2) Communication systems
(3) Safety lighting

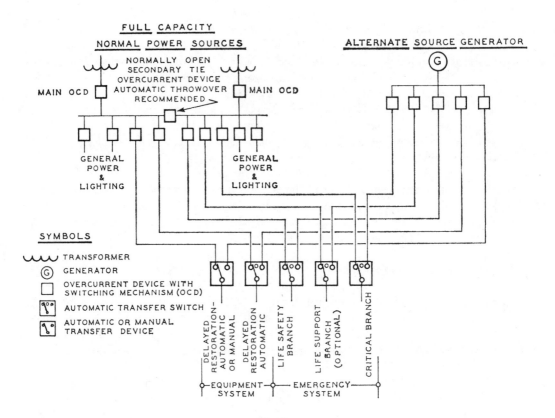

**Fig 5
Typical Hospital Wiring Arrangement
(ANSI/NFPA 99-1984 [10])**

Periodic tests and maintenance are essential to assure the reliability of the alternate source of power and other elements of the emergency system. Regular performance of maintenance is subject to the approval of the authority having jurisdiction. A record of procedures must be kept. Power outages in the past have shown the uselessness of having unmaintained emergency and standby systems that fail in time of need. Basic maintenance and test requirements for essential electric systems in hospitals are also included in ANSI/NFPA 99-1984 [10].

Since emergency and standby power systems are mandatory in most areas, the hospital electric system designer must review the state and city building laws and regulations. These laws often reference the NEC [9], ANSI/NFPA 99-1984 [10], and ANSI/NFPA 101-1985 [11], as well as other national standards such as the Uniform Building Code [24]. Insurance and building inspectors should also be consulted. In addition, installation, operation, and maintenance of the emergency power system is a factor in accreditation by the Joint Commission on Accreditation of Hospitals.

3.12.2.2 Power Quality Requirements. There is another power problem in hospitals that is not covered by any of the previously referenced codes and standards. It is the adverse effect of poor power quality on sensitive electronic loads. With the growing dependence on electronic equipment in medical facilities, users are finding that the "10 second criteria" no longer provides an acceptable functional capability. These loads are disturbed by transient overvoltages that are measured in microseconds and dips in steady-state voltage lasting only a few cycles of the 60 Hz wave. See 3.11 for a detailed description of typical power line disturbances and their effect on sensitive electronic equipment.

Some examples of sensitive hospital loads and the probable results of poor power quality (evidenced by power-line disturbances) are shown in Table 11.

This problem can be diminished if the manufacturer of sensitive or critical medical equipment thoughtfully designs the equipment to tolerate power-line disturbances. For example, built-in batteries can power equipment or at least maintain critical data during momentary interruptions in the normal supply. In addition, the susceptibility to transient overvoltage can be reduced if the equipment power supply design includes proper isolation and built-in transient overvoltage protection. These preventive measures by the medical equipment manufacturer are considered to be an efficient method of avoiding power quality-related problems.

Computer systems adapted to hospital applications typically do not have built-in protection from power-line disturbances. Nevertheless, just as in the business sector, computer systems will inevitably play a dominant role in hospitals. Organizations such as the Society for Computer Medicine provide a forum for identifying and putting into practice hospital computer applications. These applications have been summarized as "patient care computing" and include patient monitoring, laboratory, radiology, diagnostic support, patient data base acquisition, patient scheduling, and so on [22].

Table 11
Sensitive Hospital Loads

Sensitive Load	Probable Result of Severe Power Disturbances
ICU patient monitoring systems	Incorrect trend analysis or false alarm Lost time for restart and reprogramming Loss of data
Lab equipment blood gas analyzers	Depending on complexity, may result in extensive reprogramming and set-up lost time
Blood cell counter	Automatic power down for self-protection, which disrupts or delays test runs
Nuclear monitoring	Disruption of test and inability to retest due to patients' radiation exposure limit
X-ray/ultrasonic scanner	Varies from no effect to breakdown depending on equipment susceptibility
Hospital information systems (HIS) computers	Typical computer subject to disruption resulting in restart and reprogramming lost time and loss of memory (see 3.11)

With the increased dependence on computer systems in hospitals, more priority will certainly be placed on supplying continuous and disturbance-free power in the future. For computers and other life safety and life support systems where it is impractical for the equipment manufacturer to provide built-in solutions to power-quality problems, the user may find the need for power interface systems in the facility. Power interface systems bridge the gap between equipment susceptibilities and power-line disturbances. For overvoltages or electrical noise problems, isolation transformers or transient suppressors may be an effective deterrent when installed in the building power source. Required stored energy to ride-through low voltage and momentary interruptions may be obtained from a power-buffering motor-generator set or, when the need for power continuity warrants, an uninterruptible power supply (battery system). These interface systems are addressed in Chapter 4.

A large health care facility involved in extensive life-support-related functions will probably require a combination of both emergency power and power interface systems. In this case, a central system approach to satisfy special power requirements may be more cost effective than one that addresses individual equipment separately.

It is strongly recommended that the user with an apparent power quality problem consult and secure the services of qualified engineers familiar with the unique power quality requirements of the facility.

3.12.3 Other Critical Life Systems. Other critical life systems not necessarily unique to hospitals are subject to similar special laws and regulations regarding emergency power. Some examples follow:

(1) Controls for pressure vessels, such as boilers or ovens, where a failure may result in a life endangering explosion or fire

(2) Air supply systems for persons in a closed area

(3) Fire pumps, alarms, and systems

(4) Communications systems in hazardous areas

(5) Industrial processes in which the interruption of power would create a hazard to life

To meet these or other specialized emergency power requirements, the assistance of an engineer who is experienced in the particular problem area should be obtained.

3.13 Communication Systems

3.13.1 Description. Communication systems are those facilities that require electric power for verbal, written, or facsimile transmission and reception. Common systems of this type are

(1) Telephone

(2) Teletypewriter

(3) Paging

(4) Radio

(5) Television

Needs of one or all of the above communication systems during a power failure may well justify the cost of one or more emergency and standby power systems, possibly in conjunction with any other critical loads.

3.13.2 Commonly Used Auxiliary Power Systems. Battery or battery and converter equipment are practical sources of power using a float-charging system. Small engine-generators are practical and economical. In the size ranges usually required from 1–5 kW, the installed cost ranges from about $1000 per kilowatt and up, depending upon the quality of the equipment, battery life, gasoline storage problems, and amount of automatic equipment deemed to be necessary or convenient.

3.13.3 Evaluating the Need for an Auxiliary Power System. The need of an emergency or standby power system for communications should include satisfactory answers to the following questions:

(1) Will the communication equipment be required

    (a) To issue orders for an orderly shutdown of processes and equipment?

    (b) To announce instructions to personnel? (A typical announcement might be to wait by the machines or to check out for the duration of the shift.)

    (c) To call for help, issue warnings, and coordinate the work should there be a fire, civil disturbance, vandalism, or other threat to personnel safety or plant security?

(2) How will vital messages be received or sent for remote plants concerning production, inventory, or sales changes?

(3) How will key personnel be found, or instructed; and how will these persons report conditions to a central responsible source of control?

Many other questions may be asked, but the maintaining of communications under emergency conditions will save vital time and expedite the return to normal conditions with reduced confusion.

For industrial plants, the telephone system is usually powered both regularly and on a standby and emergency basis by the telephone company, normally by batteries or by standby generation. In some plants there is a separate in-plant telephone system that may be powered by batteries under a float charge that will maintain the communication system for several hours during a prolonged power interruption. The user should check to see that the system will function for the required time with loss of normal power.

Usually various bells, horns, and other call devices connected to the telephone system are powered by the lighting circuits and will stop functioning if there is a power interruption, even though the receiver and transmitter work. If required, these should be wired to the backup power system.

Teletype equipment functions much the same as the telephone system so that the signal may be present, but the printout device will not operate without local power.

Paging systems in plants are frequently extensive, using many hundreds of watts of audio power and several kilowatts of 120 V power. The need to use the system during a power interruption should not be overlooked.

Radio systems are common in industrial plants. While the mobile units for personnel as well as in-plant and out-of-plant car and truck units are generally self-powered by batteries, the main base station usually is connected to the nearest commercial power source. While mobile units may or may not be able to speak to each other, the system as a whole usually stops functioning if the base station power fails. Consideration of emergency and standby power for this base station should not be overlooked.

In some plants and commercial buildings, paging or broadcasting is done by dialing through the telephone system. In this case, both the telephone and paging or radio systems should be supplied with emergency power.

3.14 Signal Circuits

3.14.1 Description. A signal circuit supplies power to a device that gives a recognizable signal. Such devices include bells, buzzers, code-calling equipment, lights, horns, sirens, and many other devices.

3.14.2 Signal Circuits in Health Care Facilities. Signal circuits in medical buildings that should be provided with continuous emergency power within 10 s (ANSI/NFPA 101-1985 [11]) include

(1) Fire alarm systems:
Manually initiated
Automatic fire detection
Water flow alarm devices used with sprinkler systems
(2) Alarms required for systems used for piping of nonflammable medical gas
(3) Paging system
(4) Nurse's station signaling system from patient areas
(5) Alarm systems attached to equipment required to operate for the safety of major apparatus
(6) Signaling equipment for elevators in buildings of more than four stories

3.14.3 Signal Circuits in Industrial and Commercial Buildings. Signal circuits for commercial buildings and industrial plants that may require continuous emergency power within 1 min include

(1) Fire alarm system
(2) Watchman's tour system
(3) Elevator signal system
(4) Door signals (into restricted areas such as boiler rooms and laboratories with electric door locks)
(5) Liquid level, pressure, and temperature indications

3.14.4 Types of Auxiliary Power Systems. The emergency supply for the signal circuits can be engine-driven generators, multiple utility services, or floating battery systems with auxiliary power. Most signal circuits operate down to 70% rated voltage and therefore require no special voltage-sensing relays on the transfer device.

It is recommended that an emergency source of electric power be supplied for every part of a fire alarm and security system. A local battery supply on float charge, close to the power need, in continuous service is very reliable. A usually acceptable substitute is an automatic transfer to a battery system when prime power fails.

Signal circuits are usually a small electrical burden and integral part of a total load that also requires an emergency source. Therefore, the selection of emergency system and hardware generally depends upon the requirement of other related loads.

3.15 References. The following publications shall be used in conjunction with this chapter:

[1] ANSI C84.1-1982, American National Standard Voltage Ratings for Electric Power Systems and Equipment.[4]

[2] ANSI/IEEE Std 100-1984, IEEE Standard Dictionary of Electrical and Electronics Terms.

[3] ANSI/IEEE Std 141-1986, IEEE Recommended Practice for Electric Power Distribution for Industrial Plants.

[4] ANSI/IEEE Std 241-1983, IEEE Recommended Practice for Electric Power Systems in Commercial Buildings.

[5] ANSI/IEEE Std 493-1980, IEEE Recommended Practice for the Design of Reliable Industrial and Commercial Power Systems.

[6] ANSI/IEEE Std 500-1984, IEEE Guide to the Collection and Presentation of Electrical, Electronic, and Sensing Component and Mechanical Equipment Reliability Data for Nuclear Power Generating Stations.

[7] ANSI/NFPA 30-1984, Flammable and Combustible Liquids Code.[5]

[8] ANSI/NFPA 56F-1983, Nonflammable Medical Gas Systems.

[9] ANSI/NFPA 70-1987, National Electrical Code.

[10] ANSI/NFPA 99-1984, Health Care Facilities Code.

[11] ANSI/NFPA 101-1985, Life Safety Code.

[12] ANSI/NFPA 110-1985, Emergency and Standby Power Systems.

[13] ANSI/UL 924-1983, Safety Standard for Emergency Lighting and Power Equipment.[6]

[14] EGSA 101E-1984, Glossary of Terms — Electrical.[7]

[15] EGSA 101M-1984, Glossary of Terms — Mechanical.

[16] California Administrative Code, Title 24, Part 3, Basic Electrical Regulations, Article E700, Emergency Systems, Document Section, Sacramento, CA.[8]

---

[4] ANSI publications can be obtained from the Sales Department, American National Standards Institute, 1430 Broadway, New York, NY 10018.

[5] ANSI/NFPA publications can be obtained from the Sales Department, American National Standards Institute, 1430 Broadway, New York, NY 10018, or from Publication Sales, National Fire Protection Association, Batterymarch Park, Quincy, MA 02269.

[6] ANSI/UL publications can be obtained from the Sales Department, American National Standards Institute, 1430 Broadway, New York, NY 10018, or from Publication Stock, Underwriters Laboratories, Inc, 333 Pfingsten Rd, Northbrook, IL 60020.

[7] EGSA publications can be obtained from EGSA, PO Box 9257, Coral Springs, FL 33065.

[8] This document can be obtained from the Department of General Services, PO Box 1015, North Highlands, CA 95660.

[17] CASTENSCHIOLD, R. Criteria for Rating and Application of Automatic Transfer Switches. *IEEE Conference Record of 1970 Industrial and Commercial Power Systems and Electric Space Heating and Air-Conditioning Joint Technical Conference*, IEEE 70C8-IGA, pp 11-16.

[18] FUNK, D. G. and CARROLL, J. M. *Standard Handbook for Electrical Engineers*, 10th ed. New York: McGraw-Hill, 1968, pp 20-8, 20-9.

[19] GILBERT, M. M. What the Chemical Industry Can Do to Minimize Effects from Electrical Disturbances. Presented at the AIEE and ASME National Power Conference, Paper CP 60-1161, Philadelphia, PA, Sept 21-23, 1960.

[20] IEEE Committee Report. *Protection Fundamentals for Low-Voltage Electrical Distribution Systems in Commercial Buildings*. IEEE JH 2112-1, 1974, pp 19, 20.

[21] KAUFMAN, J. E., Ed. *IES Lighting Handbook*. New York: Illuminating Engineering Society, 1972, sec 14, pp 14-10-14-12.

[22] *Readers Digest Almanac*. Pleasantville, NY: The Readers Digest Association, 1972, p 790.

[23] Standard Building Code, Southern Building Code Congress International.[9]

[24] Uniform Building Code, International Congress of Building Officials.[10]

[25] Williams-Steiger Occupational Safety and Health Act of 1970 (84 Stat 1593, 1600; 29 USC 656, 657), Public Law no 91-596, Chapter XVII of Title 29 of The Code of Federal Regulations, established on April 13, 1970 (36 FR 7006) as amended by adding thereto a new part 1910, Washington, DC, Department of Labor. Amendments to Williams-Steiger Occupational Safety and Health Act of 1970, as published in the Federal Register, Washington, DC, Office of the Federal Register, National Archives and Records Service, General Services Administration.[11]

3.16 Bibliography

[B1] Beating the Blackouts. *The Wall Street Journal*, July 21, 1970.

[B2] BEEMAN, D. L., Ed. *Industrial Power Systems Handbook*. New York: McGraw-Hill, 1955.

[B3] BURCH, B. F., Jr. Protection of Computers against Transients, Interruptions, and Outages. Presented at the 1967 IEEE Industry and General Applications Group Annual Meeting, Pittsburgh, PA, Oct 4, 1967.

---

[9] This document can be obtained from the Southern Building Code Congress International, 900 Montclair Rd, Birmingham, AL 35213.

[10] This document can be obtained from the International Conference of Building Officials, South Workman Mill Rd, Whittier, CA 90601.

[11] This document can be obtained from the Occupational Safety and Health Administration, 1515 Broadway, New York, NY 10036.

[B4] FERENCY, N. Is a UPS Really Necessary? *Electronic Products Magazine*, Apr 17, 1972, pp 126–127.

[B5] FISCHER, E. I. Emergency Power Facilities for a Research Laboratory. *Conference Record of the 1966 IEEE Industry and General Applications Group Annual Meeting*, IEEE 34-C36, pp 305–324.

[B6] HEISING, C. R. and DUNKI-JACOBS, J. R. Application of Reliability Concepts to Industrial Power Systems. *Conference Record of the 1972 IEEE Industrial Applications Society Annual Meeting*, IEEE 72CH0685-81A, pp 289–295.

[B7] JENKIN, M. A. Functions of Patient Care Computing, *Medical Instrumentation*, vol 12, July–Aug 1978.

[B8] KATZ, E. G. Evaluation of Hospital Essential Electrical Systems. *Fire Journal*, vol 62, Nov 1968.

[B9] KEY, T. S. Diagnosing Power Quality Related Computer Problems. *IEEE ICPS Conference Record*, IEEE 78CH1302-91A, June 1978, pp 48–56.

[B10] KNIGHT, R. L. and YUEN, M. H. The Uninterruptible Power Evolution — Are Our Problems Solved? *Conference Record of the 1973 IEEE Industry Applications Society Annual Meeting*, IEEE 73CH0763-31A, pp 481–485.

[B11] MILLER, N. A. Noninterruptible Power Supplies for Essential Control Systems in Power Plants. Presented at the IEEE IAS–IEC Group Meeting on Heating, Chicago Chapter, June 7, 1972.

[B12] OLIVER, R. L. Plant Engineering at RCA, *Plant Engineering*, Apr 1969.

[B13] The On-Site Power Market. *Electrical Construction and Maintenance*, Jan 1971, pp 57–71.

[B14] TUCKER, R. Line Voltage Regulators, Less Costly Alternative to UPS. *Electrical Consultant*, Aug 1974.

Chapter 4
Generator and Electric Utility Systems

4.1 Guidelines for Use. When a user experiences an equipment problem due to failure of the electric power supply, he must either live with the problem, change his equipment or system to perform satisfactorily during a failure of the existing power supply, or alter the supply to prevent potential failures. In many cases, the correct decision is to change the equipment or system, but this chapter does not examine these cases.

Once the electric power user's study has shown that the correct approach is to alter or supplement the power supply source, a study should be undertaken to determine the proper systems and hardware that will meet the need at the lowest cost for the electric power required.

This chapter describes combinations of systems and hardware that will overcome the following types of electric power failures with good reliability:

(1) Long-time interruption (hours)
(2) Medium-time interruption (minutes)
(3) Short-time interruption (seconds)
(4) Over- or undervoltage
(5) Over- or underfrequency

Emergency power systems are of two basic types: (1) an electric power source separate from the prime source of power, operating in parallel, that maintains power to the critical loads should the prime source fail; or (2) an available reliable power source to which critical loads are rapidly switched automatically when the prime source of power fails.

Standby power systems are made up of the following main components:

(1) An alternative reliable source of electric energy separate from the prime power source

(2) Starting and regulating control if on-site standby generation is selected as the source

(3) Controls that transfer loads from the prime or emergency power source to the standby source

99

It is prudent for the user to establish his practical need from the previous chapters of this publication before specifying and purchasing equipment, since costs will rise as the following requirements are specified for systems and hardware:

(1) Longer equipment life
(2) Increased capacity
(3) Closer frequency regulation
(4) Closer voltage regulation
(5) Freedom from voltage or frequency transients
(6) Increased availability
(7) Increased reliability
(8) Increased temporary overload capability
(9) Quiet operation
(10) Safety from fuel hazards
(11) Pollution-free operation
(12) Freedom from harmonics
(13) Close voltage and frequency regulation with wider range rapid load changes

Consideration should be given to load growth. Future power requirements frequently need to be connected to the emergency and standby system. It may be desirable to add additional existing power loads to the more reliable power bus as soon as the advantages are realized in practice. If additional capacity cannot be justified initially, the equipment and system should be selected and designed for future economic expansion compatible with the initial installation.

Operating costs of the systems and hardware are usually secondary to meeting the need, but should be included as a factor in the selection. These include cost of fuel, inspection frequency, ease of maintenance, frequency of testing, cost of parts, and taxes.

Installation quality should be high to prevent losing the reliability of electric power designed into the system and purchasd in the hardware [18].[12] One should guard against introducing voltage transients into the emergency and standby power distribution system. Satisfactory voltage levels should be maintained under all loading conditions.

For industrial plants, electric systems should conform to ANSI/IEEE Std 141-1986 [4]. For commercial buildings, electric systems should conform to ANSI/IEEE Std 241-1983 [6]. Grounding practices should follow the recommendations in ANSI/IEEE Std 142-1982 [5] and Chapter 7 of this book. Additional shielding, bonding, grounding, and even filtering may be required to maintain the quality of the emergency and standby power supply.

When energized sources of power are available for emergency use, a light should be included in the system to show that the source is energized and an alarm, to signal loss of available power. These should be located where responsible persons

---

[12] The numbers in brackets correspond to those of the references listed at the end of this chapter; when preceded by B, they correspond to the bibliography at the end of this chapter.

can take action, should an alarm sound. An alternate utility source (4.3) or battery system is typical of line sources normally requiring these signals.

Users should take these additional steps to assure performance reliability:

(1) Establish regular inspections using a check sheet and recording exceptions

(2) Perform regular preventive maintenance and repair items of exceptions found during the inspections

(3) Set up a trial at regular intervals simulating a power failure but timed so as not to encounter hazards or losses should the system not operate as anticipated

There are mathematical methods for quantitatively determining the reliability of an emergency or standby power system. Once the need is established, it may be advisable to calculate the system reliability, especially for emergency systems involving possible injury or loss of life. One such method is detailed in Sawyer [19] and Heising and Johnston [17].

4.2 Engine-Driven Generators

4.2.1 Introduction. These units are *work horses* that fulfill the need for emergency and standby power. They are available from small 1 kVA units to those of several thousand kVA. When properly maintained and kept warm, they dependably come on line within 8–15 s. In addition to providing emergency power, engine-driven generators are also used for handling peak loads and are sometimes used as the preferred source of power.

They fill the need of back-up power for uninterruptible power systems. Where well-regulated systems, free from voltage, frequency, or harmonic disturbances, are required, such as for computer operations, a *buffer* may be needed between the critical load and the engine-driven generators.

4.2.2 Diesel-Engine Generators. A typical diesel engine-driven generator rated 500 kW is shown in Fig 6. Typical ratings of engine-driven generator sets are given in Table 12. Individual models by various companies may be different. Lower speed units are heavier and more costly, but are more suitable for continuous duty.

Diesel engines are somewhat more costly and heavier in smaller sizes, but are rugged and dependable. The fire and explosion hazard is considerably lower than for gasoline engines. Sizes vary from about 2.5 kW to several MW.

4.2.3 Gasoline-Engine Generators. Gasoline engines may be furnished for installations up to about 100 kW output. They start rapidly and are low in initial cost as compared to diesel engines. Disadvantages include a higher operating cost, a greater hazard due to the storing and handling of gasoline, short storage life of the fuel, and generally a lower mean-time between overhaul. The short fuel storage life restricts gasoline engines' use for emergency standby.

4.2.4 Gas-Engine Generators. Natural gas and liquid petroleum (LP) gas engines rank with gasoline engines in cost and are available up to about 600 kW and higher. They provide quick starting after long shutdown periods because of the fresh fuel supply. Engine life is longer with reduced maintenance because of the clean burning of natural gas. However, consideration should be given to the possibility of both the electric utility and the natural gas supply being concurrently unavailable. Unless engine compression ratios are increased, engines lose approxi-

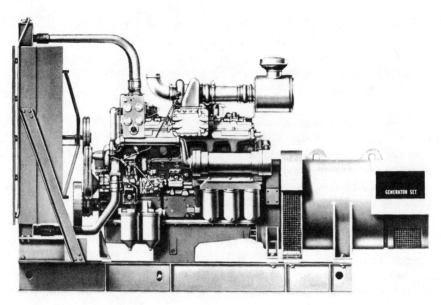

Fig 6
Typical Diesel Engine-Driven Generator

Table 12
Typical Ratings of Engine-Driven Generator Sets

Nominal Rating (kW)	Prime Power Rating (kW)	Standby Rating (kW)	Power Factor	Prime Mover			Speed (r/min)
				Gasoline	Diesel	Natural Gas/ LP Gas	
5	5	5	1.0	×		×	3600
10	10	10	1.0	×		×	1800
25	25	30	0.8		×	×	1800
100	90	100	0.8		×	×	1800
250	225	250	0.8		×		1800
750	665	750	0.8		×		1800
1000	875	1000	0.8		×		1800

mately 15% in power when operated on natural gas, compared to gasoline. Considerations in selecting natural or LP gas-fueled engines are the availability and dependability of the fuel supply, especially in an emergency situation. Reference should be made to ANSI/NFPA 70-1987 [10] (National Electrical Code [NEC]), Articles 700-12 and 701-10, to determine whether an on-site fuel supply is required.

4.2.5 Derating Requirements. As noted Fig 29 in 4.4.3, altitude will cause a serious derating of the prime mover to deliver the torque required for the full generator output. Oversizing of the engine is a must for higher altitudes. The generators are less critical in output capacity if adequate cooling air is supplied to carry away the heat generated by the losses.

A general rule for derating engine power loss with altitude increase is to derate about 4% for each 1000 ft increase in altitude. Turbo-charged engines usually do not need to be derated below a specified minimum altitude, typically 2500-5000 ft above sea level. Also, an average derating factor for high ambient temperature is 1% for each 10 °F above 60 °F. Temperature derating is not considered as important as altitude derating.

4.2.6 Multiple Engine-Generator Set Systems. Automatic starting of multiple units and automatic synchronizing controls are available and practical for multiple-unit installations. Advantages of several smaller units over one large unit should be considered since emergency and standby power can be available while one unit is being maintained or overhauled. Starting is usually reliable; especially if the units are warm and maintained by regular exercising, the likelihood of all of the units not starting is extremely low as compared to a single unit.

Smaller units also allow the *building block* concept. As capital is made available and the need of increased capacity grows, additional units of identical size and type may be added, thus simplifying the parts, maintenance, and training problems. The trend toward larger emergency power supplies also justifies the use of multiple sets to provide the additional power. Figs 7, 8, 9, and 10 illustrate typical multiple-set systems.

One very important consideration in the selection of smaller versus larger units is the service to which they will be subjected. Smaller units are nearly all higher speed (1800 r/min) engine-driven sets, and if these units are to be run continuously for long periods of time, the engine should be evaluated very thoroughly. Refer to Table 12 for the comparison between prime power and standby ratings. Units may be operated at the standby rating for the duration of a power outage but should not be used at that rating for continuous operation. Generators for standby use are frequently operated at higher output and temperature rise than are those for continuous use.

4.2.7 Construction and Controls The basic electrical components are the engine-generator set and associated meters, controls, and switchgear. Most installations include a single generator set designed to serve either all the normal electrical needs of a building or a limited emergency circuit. Sometimes the system includes two or more generators of different types and sizes serving different types of loads. Also, two or more generators may be operating in parallel to serve the same load.

4.2.8 Typical Engine-Generator Systems. This information is designed to help in selecting the electrical components of a generator installation. No attempt has been made to cover every situation that might arise.

The diagrams in this chapter show a circuit breaker at each generator set. The NEC [10], Article 445, permits a number of forms of generator protection including inherent protection. Therefore, a circuit breaker or other form of overcurrent protection may or may not be used. For an in-depth discussion of generator protection, see Chapter 6.

In Figs 7, 8, 9, 10, and 11, the following abbreviations are used:

ATD = Automatic transfer device (automatic transfer switch or electrically operated circuit breaker)

CB = Circuit breaker

EG = Engine-driven generator set

LDC = Load-dumping contactor, electrically operated, mechanically held

PC = Paralleling contactor

Figure 7 shows a standby power system in which, if power fails from the normal source, both engines automatically start. The first generator to reach operating voltage and frequency will actuate load dumping circuits and cause the remaining

Fig 7
Two Engine-Generator Sets Operating in Parallel

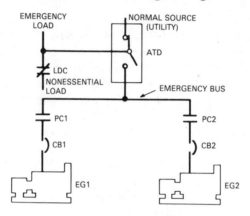

Fig 8
Peaking Power Control System

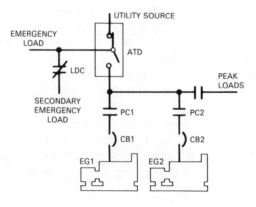

load to transfer to this generator. When the second generator is in synchronism, it will be paralleled automatically with the first. After the generators are paralleled, all or part of the dumped load is reconnected if the standby capacity is adequate.

If one generator fails, it is immediately disconnected. A proportionate share of the load is dumped to reduce the load to where the remaining generator can handle it. When the failed generator is reinstated, the dumped load is reconnected. When the normal source is restored, the load is retransferred and the generators are automatically disconnected and shut down.

With the system shown in Fig 8, idle standby generator sets can perform a secondary function by helping to supply power for peak loads. Depending on the load requirements, this system starts one unit or more to supply peak loads while the utility service supplies the emergency circuits. When the second generator is in synchronism, it will be paralleled automatically with the first. If the utility service fails, the peak loads are automatically disconnected and the generators pick up the emergency loads through the transfer device. Use of the standby generator(s) for peak shaving may increase wear and consequently shorten the overall time between overhauls. With proper maintenance, however, the reliability of the system is not hindered. Peak shaving use may require a different air quality permit.

Figure 9 shows a standby power system where there is a split emergency load with one load being more critical than the other.

When the prime source fails, both generators start. If Load 1 is the preferential load, the generator that reaches operating speed first is put on the line by Automatic Transfer Device 3 to feed Load 1 through Automatic Transfer Device 1. When the other generator reaches operating speed, it then feeds Load 2. If the generator feeding Load 1 fails at any time, the other generator will be transferred from Load 2 and take over Load 1. When the prime source is restored, both loads are retransferred to the normal source and the generators shut down.

Fig 9
Three-Source Priority Load Selection System

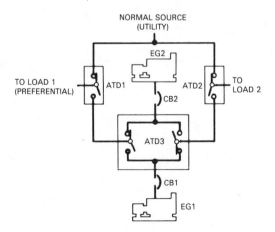

The system shown in Fig 10 provides switching and control of utility and on-site power. Two on-site buses are provided: (1) an on-site power bus (preferred) supplies continuous power for computer or other essential loads, and (2) an emergency bus (secondary) supplies on-site generator power to emergency loads through an automatic transfer device if the utility service fails.

In normal operation, one of the generators is selected to supply continuous power to the preferred bus (in Fig 10, EG1). Simplified semiautomatic synchronizing and paralleling controls permit any of the idle generators to be started and paralleled with the running generator to change generators without load interruption. Anticipatory failure circuits for low oil pressure and high coolant temperature permit load transfer to a new generator without load interruption. However, if the generator enters a critical failure mode, transfer to a new generator is made automatically with load interruption.

Many loads, such as lighting, fire alarms, heating, and air-conditioning, are fed by the utility service through the transfer device. If the utility fails, idle generators are automatically started and assume these loads through the automatic transfer device.

Fig 10
Combination On-Site Power and Emergency Transfer System

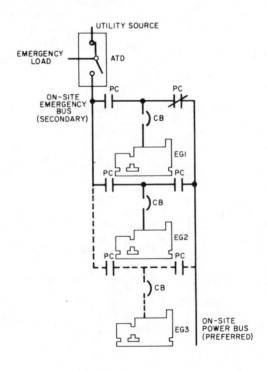

Figure 11 is similar to Fig 8 wherein idle standby generator sets can perform a secondary function by helping to supply power for peak loads. Depending on the load requirements, this system starts one or both generators to feed the peak load by switching ATD2 while the utility service continues to supply the emergency and prime loads. The second generator is paralleled automatically with the first.

If the utility service fails, the emergency and prime loads are automatically transferred to the emergency generators. Depending on generator capacity, the peaking load may be left on, or CB5 may be dropped for the emergency.

4.2.9 Special Considerations. Unusual conditions of altitude, ambient temperature, or ventilation may require either a larger generator to hold down winding temperatures or special insulation to withstand higher temperatures. Generators operating in the tropics are apt to encounter excessive moisture, high temperature, fungus, vermin, etc, and may require special tropical insulation and space heaters to keep the windings dry and the insulation from deteriorating.

4.2.10 Engine-Generator Set Rating. For some buildings the maximum continuous generator load will be the total load when all equipment in the building is operating. For others, it may be more practical and economical to set up an emergency circuit or circuits so that only certain essential lights and equipment, and perhaps just one elevator, can be operated when the load is on the standby generator.

4.2.11 Motor-Starting Considerations. If the maximum momentary voltage dip that is acceptable in the circuit is known, it is possible to select the size of the engine-driven generator that will be able to start given sized motors without exceeding the voltage dip. If it is possible that two motors can start together, the sum of their horsepower ratings should be used as a basis for estimating motor starting requirements or controls provided for separate starting.

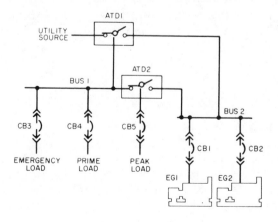

**Fig 11
Dual Engine-Generator Standby System**

Engines driving generators should be sized to handle the continuous kilowatt load to be supplied to the load plus motor-starting requirements and the generator losses at the number of kilowatts selected.

In sizing the engine-generators for motor starting, the *locked-rotor* or inrush kilovolt-ampere rating of the motors should be used. Manufacturers' data can usually be obtained, giving the maximum rating in kilovolt-amperes of the engine generator, as well as its continuous rating. The maximum rating would be the maximum number of short-duration kilovolt-amperes available for motor-starting duty without exceeding a specified voltage dip. Motor-starting load has a very low power factor that should be considered in calculating the voltage dip. One other important consideration is the effect of the generator voltage dip on the motor-starting torque. The starting torque is proportional to the kilovolt-ampere input to the motor, but since a voltage dip to 70% of rated voltage results in a reduction of power to approximately 50% of the number of kilovolt-amperes into the stalled rotor and thus a 50% reduction of motor-starting torque, problems could result in starting motors under load, unless this factor is taken into consideration. Generator set manufacturers are usually willing to furnish a guide for calculating motor-starting effects. A rule of thumb of 0.5 hp/kW is frequently used; however, the final decision should be based upon the manufacturer's data. When motor-starting kilovolt-amperes or kilowatts exceed the rated values of the generator set, the effects of the resulting voltage and frequency deviations on equipment other than the motor being started should be evaluated (that is, motor starters, relays, computers, communication equipment, etc).

Generators are usually sized for the maximum continuous kilovolt-ampere demand. Should there be unusually high inertia loads to start without benefit of reduced-voltage starting, or if voltage and frequency regulation other than specified cannot be tolerated during the startup period, a higher rated generator may be required.

4.2.12 Load Transient Considerations. A voltage regulator with sufficient response is required to minimize voltage sags or surges after load transients. The engine-generator set should be of sufficient capacity and design capability to minimize the effect of load transients. Many industrial applications can tolerate large voltage sags (usually down to 80%, but as low as 65% in special cases) as long as they are not so low as to cause motor contactors to drop out or automatic brakes to set. Solid-state controls and computers may be affected.

4.2.13 Manual Systems. Manually controlled standby service is the simplest and lowest cost arrangement and may be satisfactory where an attendant is on duty at all times and where automatic starting and transfer of the load is not a critical requirement.

4.2.14 Automatic Systems. In order for engine-driven generators to provide automatic emergency power, the system should also include automatic engine starting controls, automatic battery charger, and an automatic transfer device. In most applications, the utility source is the normal source and the engine-generator set provides emergency power when utility power is interrupted or its characteristics are unsatisfactory. The utility power supply is monitored and engine starting is automatically initiated once there is a failure or severe voltage or frequency reduction in the normal supply. The load is automatically transferred as soon as the standby generator stabilizes at rated voltage and speed. Upon restoration of normal supply, the transfer device automatically retransfers the load and initiates engine shutdown.

4.2.15 Automatic Transfer Devices. Transfer equipment for use with engine-generator sets is similar to that used with multiple-utility systems, except for the addition of auxiliary contacts that close when the normal source is interrupted. These auxiliary contacts initiate the starting and stopping of the engine-driven generator. The automatic transfer device may also include accessories for automatic exercising of the engine-generator set and a 5 min unloaded running time before shutdown. For additional information on automatic transfer devices, refer to 4.3.

4.2.16 Engine-Generator Set Reliability. To keep the engine in good condition, whenever it starts it should run for a sufficiently long time so that all parts reach their normal operating temperatures. In case emergency power is called for only briefly, it is desirable that the engine continue to run for about 15 min after normal power has been restored. A programmed control should be added that starts the engine once a week and operates it for a set period of time under load.

Reliability and satisfaction of the standby power supply partially depends on the engine generator installations. The following key points should be considered:

(1) Systems dependent upon a municipal water supply are not dependable because the supply can be interrupted in an emergency. A radiator cooling system or a heat exchanger cooled system with an independent cooling water supply should be specified.

(2) Antifreeze protection may be required. A room heater, electric immersion heater, or steam jacket improves starting reliability and solves the antifreeze problem at the same time. A loss-of-heat alarm may be required.

(3) A silencer on the intake and exhaust may be required by law and will reduce noise levels.

(4) Fuel supply systems must meet local laws, regulations, and insurance requirements. The capacity stored depends on available guaranteed quick delivery replacement, including Sundays and holidays and under all weather conditions.

(5) Fuel supplies stored underground are desirable. Above ground, antifreeze protection may be required.

(6) Gasoline and diesel fuels deteriorate if they stand unused for a period of several months. Normal testing and running on line may be used to keep fuel fresh. Inhibitors may be added to the fuel. Fuel used for several purposes will tend to be fresh.

(7) Engine vibration transferred to the building should be dampened by rubber pads or springs, and flexible couplings should be used on fuel, exhaust, water, and conduit lines.

(8) Equipment mounting should conform to local seismic requirements.

(9) Starting aids are of three types, of which (c) is frequently preferred:
  (a) Air heated before it reaches the cylinders
  (b) Volatile starting fluid injected into the engine air
  (c) Engine block maintained warm

4.2.17 Air Supply and Exhaust. Exhaust piping inside the building should be covered with gas-tight insulation to protect personnel and to reduce room temperature. The exhaust piping must be of sufficient diameter to avoid exhaust back pressure. Consideration should be given to dissipating the exhaust away from air intakes and minimizing air pollution.

Some means of providing free flow of fresh air into the generator room is necessary to keep the atmosphere comfortable for personnel and to make clean, cool air available to the engine. The most effective and lowest-cost method is usually to use a *pusher*-type fan on the radiator and connect the radiator to the outside through a duct. The intake opening should be approximately 25–50% larger than the duct.

4.2.18 Noise Reduction. Vibration must frequently be isolated from structures to reduce noise. Noise-reducing mufflers are rated according to their degree of silencing by such terms as *industrial*, *residential*, or *critical*, and are usually required to meet noise standards. An intake muffler is not usually installed since an engine normally is supplied with at least one intake air cleaner that also serves as an intake silencer.

4.2.19 Fuel Systems. To simplify the fuel supply system, the fuel tank should be as close to the engine as possible. When gasoline or LP fuel is used, it normally cannot be stored in the same room with the engine because there is a danger of fire or fumes. However, when diesel fuel is used, it can sometimes be stored in the same room as the engine. If a remote tank is used, a transfer tank located near the engine is recommended. The building code or fire insurance regulations should be checked to determine whether the fuel storage tank may be located beside the generator set, in an adjacent room, outside, or underground.

An engine equipped to operate on gasoline, LP gas, or diesel fuel stored on-site is a self-contained system that does not depend on outside municipal or utility services. It is dependable and affords independent standby protection.

4.2.20 Governors and Regulation. Governors are of two types, droop and isochronous. With a droop-type governor, the engine's speed is slightly higher at light loads than at heavy loads, while an isochronous governor maintains the same steady speed at any load up to full load:

$$\text{speed regulation} = \frac{\text{no load r/min} - \text{full load r/min}}{\text{full load r/min}} \cdot 100\%$$

A typical speed regulation for a droop-type generator is 3%. Thus if speed and frequency at full load are 1800 r/min and 60 Hz, at no load they will be about 1854 r/min and 61.8 Hz. A droop-type governor usually is set so that it holds the desired nominal speed at full load.

Under steady load, frequency tends to vary slightly above and below the normal frequency setting of the governor. The extent of this variation is a measure of the stability of the governor. An isochronous governor should maintain frequency regulation within ± ¼%.

When load is added or removed, speed and frequency dip or rise momentarily, usually 1–3 s, before the governor causes the engine to settle at a steady speed at the new load. For generators operating in parallel with a primary source of power, the governor may be arranged to automatically switch from droop to isochronous mode upon loss of the primary source.

4.2.21 Starting Methods. Most engine-generator sets utilize a battery-powered electric motor for starting the engine. A pneumatic or hydraulic system normally is used on large units where battery starting is impracticable.

4.2.22 Battery Charging. In addition to the battery-charging generator on the set, a separate automatic battery charger is recommended for maintaining battery charge when the generator is not running. Failure to keep the battery properly charged is the greatest cause of emergency system failure.

4.2.23 Advantages and Disadvantages of Diesel-Driven Generators. In evaluating the merits of diesel engine versus gas turbine prime movers, the following advantages and disadvantages of each should be considered.

(1) *Fuel Supply.* Gas turbines and diesel engines can generally burn the same fuels (kerosene through #2 diesel).

(2) *Starting.* Where the application requires acceptance of 100% load in 10 s, diesel generator sets can be provided that can meet this requirement. Most gas turbine generators sets require more than 30 s.

(3) *Noise.* The gas turbine operates quieter and has less vibration than the diesel engine.

(4) *Ratings.* The gas turbine is not readily available in sizes less than 500 kW, while diesel engine units range from 15 kW or lower.

(5) *Cooling.* Diesel engines in the larger sizes normally require water cooling, while gas turbines are normally air-cooled.

(6) *Installation.* Gas turbines are considerably lighter and smaller in size. Turbines also require less total cooling and combustion air and produce minimal vibration. Installation costs are normally less and rooftop applications are more feasible.

(7) *Cost.* First cost for diesel engines is lower than gas turbines, but overall installed cost sometimes becomes comparable due to the lower installation costs of the gas turbines.

(8) *Exercising.* The cyclic operating requirements under load are more rigid for diesel units than for gas turbines.

(9) *Maintenance.* The gas turbine is a more simple machine than a diesel engine. However, repair service for a diesel engine is generally more readily available than for a gas turbine.

(10) *Efficiency.* The diesel engine operates more efficiently than a gas turbine under full load. However, the reduced exercising requirements for the gas turbine normally make the turbine the lower fuel consumer in standby applications.

(11) *Frequency Response.* The gas turbine generator is superior in full load transient frequency response.

4.2.24 Additional Information. Engine generator specifications are treated in greater detail in EGSA 101P-1985 [16].

4.3 Multiple Utility Services

4.3.1 Introduction. Multiple utility services may be used as an emergency or standby source of power. Required is an additional utility service from a separate source and the required switching equipment. Figure 12 shows automatic transfer

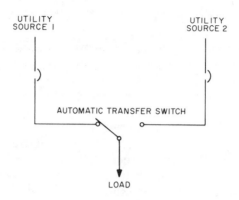

Fig 12
Two-Utility-Source System Using One Automatic Transfer Switch

between two low-voltage utility supplies. Utility source 1 is the normal power line and utility source 2 is a separate utility supply providing emergency power. Both circuit breakers are normally closed. The load must be able to tolerate the few cycles of interruption while the automatic transfer device operates.

4.3.2 Closed-Transition Transfer. If the utility will permit the two sources of supply to be connected together momentarily, the transfer device may be provided with controls for both open (normal supply opened before the emergency supply is closed) and closed (emergency supply closed before the normal supply is opened) transition. With closed transition, the utility can notify the customer to transfer to the emergency source in order to take the normal supply out of service for maintenance and repair without the momentary interruption that occurs with open transition. Closed transition requires the sources to be synchronized with proper phase angle and phase sequence.

4.3.3 Utility Services Separation. Use of multiple utility service is economically feasible when the local utility can provide two or more service connections over separate lines and from separate supply points that are not apt to be jointly affected by system disturbances, storms, or other hazards. It has the advantage of relatively fast transfer in that there is no 5-15 s delay as there is when starting a standby engine-generator set. A separate utility supply for an emergency should not be relied upon unless total loss of power can be tolerated on rare occasions. Otherwise, use of engine-generator sets is recommended. Also, in some installations, such as hospitals, codes require on-site generators.

A no-potential alarm should be installed on the emergency supply so that the utility can be notified and emergency precautions taken if the emergency supply is lost. Additional reliability has been obtained in rare instances where the services are available from different utility companies.

4.3.4 Simple Automatic Transfer Schemes. Automatic switching equipment may consist of three circuit breakers with suitable control and interlocks, as shown in Fig 13. Circuit breakers are generally used for primary switching where the voltage exceeds 600 V. They are more expensive but safer to operate, and the use of

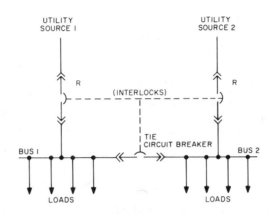

Fig 13
Two-Utility-Source System Where Any Two Circuit Breakers Can Be Closed

fuses for overcurrent protection is avoided. Relaying is provided to transfer the load automatically to either source if the other one fails, provided that circuit is energized. The supplying utility will normally designate which source is for normal use and which is for emergency. If either supply is not able to carry the entire load, provisions must be made to drop noncritical loads before transfer takes place. If the load can be taken from both services, the two R circuit breakers are closed and the tie circuit breaker is open. The three circuit breakers are interlocked to permit any two to be closed but prevent all three from being closed. The advantage of this arrangement is that the momentary transfer outage will occur only on the load supplied from the circuit that is lost. However, the supplying utility may not allow the load to be taken from both sources, especially since a more expensive totalizing meter may be required. A manual override of the interlock system should be provided so that a closed transition transfer can be made if the supplying utility wants to take either line out of service for maintenance or repair and a momentary tie is permitted.

If the supplying utility will not permit power to be taken from both sources, the control system must be arranged so that the circuit breaker on the normal source is closed, the tie circuit breaker is closed, and the emergency source circuit breaker is open. If the utility will not permit dual or totalized metering, the two sources must be connected together to provide a common metering point and then connected to the distribution switchboard. In this case the tie circuit breaker can be eliminated and the two circuit breakers act as a transfer device. Under these conditions the cost of extra circuit breakers can rarely be justified.

The arrangement shown in Fig 13 only provides protection against failure of the normal utility service. Continuity of power to critical loads can also be disrupted by

(1) an open circuit within the building (load side of the incoming service),

(2) an overload or fault tripping out a circuit, or

(3) electrical or mechanical failure of the electric power distribution system within the building.

It may be desirable to locate transfer devices close to the load and have the operation of the transfer devices independent of overcurrent protection. Multiple transfer devices of lower current rating, each supplying a part of the load, may be used rather than one transfer device for the entire load.

Availability of multiple utility service systems can be improved by adding a standby engine-generator set capable of supplying the more critical load. Such an arrangement, using multiple automatic transfer switches, is shown in Fig 14.

4.3.5 Overcurrent Protection. Caution should be exercised to assure that transfer control and operation do not in any way detract from overcurrent protec-

**Fig 14
Two Utility Sources Combined with an Engine-Generator Set
to Provide Varying Degrees of Emergency Power**

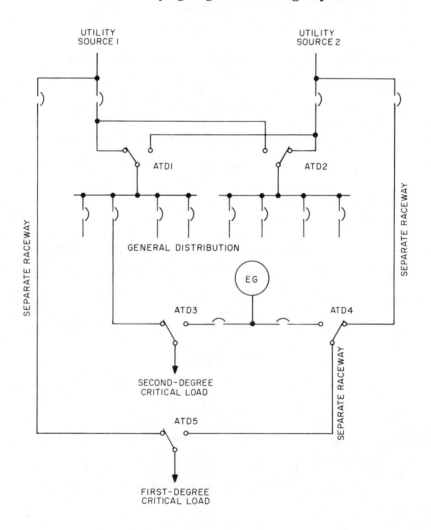

tion and vice versa. Furthermore, the transfer and overcurrent protective devices should be so arranged that means for disconnecting incoming service is conventional and readily accessible. Chapter 6 provides a more in-depth analysis of overcurrent protection.

4.3.6 Transfer Device Ratings and Accessories. The required characteristics of transfer devices should include the capabilities of

(1) closing against inrush currents without contact welding

(2) carrying full rated current continuously without overheating

(3) withstanding available short-circuit currents without contact separation

(4) properly interrupting the loads to avoid flashover between the two utility services

In addition to considering each of the above individually, it is also necessary to consider the effect each has on the other. Particular consideration should be given to coordination between automatic transfer switches and overcurrent protection. High fault currents create electromagnet forces within the contact structure of circuit breakers, which help provide fast opening and therefore minimum clearing time; however, automatic transfer switches designed to withstand high fault currents utilize these electromagnetic forces in a reverse manner to assure that the transfer switch contacts remain closed until the fault has been cleared. Contact separation, while carrying fault level current, results in arcing, melting of contact surface, and possible contact welding upon reclosure and cooling of molten metal. ANSI/UL 1008-1983 [12] does not permit contact welding after the transfer switch has been subjected to the withstand current test, which is the recommended means of verifying this capability. For these reasons, transfer switching devices should be selected from those designed and approved for the purpose.

Most transfer switches are capable of carrying 100% rated current at an ambient temperature of 40 °C. However, transfer switches incorporating integral overcurrent protective devices may be limited to a continuous load current not to exceed 80% of the switch rating. The manufacturer's specification sheets indicate whether the device is 80 or 100% rated.

Transfer switches differ from other emergency equipment in that they continuously monitor the utility source and continuously carry current to critical loads. Fault currents, repetitive switching of all types of loads, and adverse environmental conditions should not cause excessive temperature rise or detract from reliable operation.

Most transfer switches are rated for total system transfer and thus are suitable for continuous duty control of motors, electric discharge lamps, tungsten filament lamps, and electric heating equipment, provided that the tungsten lamp load does not exceed 30% of the switch rating. Some manufacturers also provide transfer switches for 100% tungsten lamp load. There are cases where transfer switches are limited to specific loads, such as resistance loads only. For these reasons, the load classification should be determined when selecting transfer switches.

Load transfer devices are available in the following forms:

(1) Automatic transfer switches are available in ratings from 30–4000 A to 600 V (Fig 15) [19].

Fig 15
Modular-Type Automatic Transfer Switch Suitable for All Classes of Load

(2) Automatic power circuit breakers consisting of two or more power circuit-breakers that are mechanically or electrically interlocked, or both, rated 600–3000 A, to 15 kV.

(3) Manual transfer switches (600 V) available in current ratings from 30–200 A.

(4) Nonautomatic transfer switches available in ratings from 30–4000 A to 600 V, manually controlled and electrically operated.

(5) Manual or electrically operated bolted pressure switches (600 V) fusible or nonfusible available from 800–6000 A.

Features and accessories depend upon the form of transfer device and may include the following:

(1) Voltage monitors that have adjustable settings for dropout (unacceptable voltage) and pickup (acceptable voltage). The dropout initiates transfer of the load to the alternate source; the pickup initiates retransfer of the load back to the

normal source. The dropout range is typically 75–95% of the setting selected for pickup. The typical pickup setting range is 85–98% of nominal voltage. Usual setting for most types of load are 85% of nominal voltage for dropout and 90% for pickup. (Dropout is usually about 95% of pickup.) Lower dropout or additional time delay may be necessary if significant voltage drop is produced by starting large motors.

(2) Test switch to simulate a power failure to provide a periodic test of the emergency source and the transfer operation.

(3) Controls for closed transition if the utility will permit the two sources to be momentarily tied together and if the switch design permits. This will allow manual transfer from the normal to the emergency feeder and back without the momentary transfer interruption.

(4) Provisions for switching the neutral conductor and thereby minimizing ground currents and simplifying ground-fault sensing. For further discussion, see Chapter 7.

(5) Controls for motor load transfer so as to avoid abnormal currents caused by the motor's residual voltage being out of phase with the voltage source to which the motor is being transferred.

(6) Circuitry to initiate starting and stopping of the engine-generator set(s) depending upon availability of the normal source of power.

(7) Time delays in the transfer switch to permit proper programming of transfer switch operation. A time-delay relay, usually adjustable from 0–6 s, is used to provide override of momentary interruptions and to initiate engine start and power transfer if the outage or voltage reduction is sustained. This feature is necessary to minimize wear on the engine-starting motor, ring gear, and other system equipment and to prevent unnecessary drain on the engine-cranking battery.

This time delay is usually set at about 1 s, but can be set for longer if circumstances warrant. If delay settings are longer, care needs to be taken to ensure that sufficient time remains to get the alternate power source on line within the time prescribed by applicable codes if the system is a required system.

After the engine has started and generator output is at proper voltage and frequency, another time-delay relay is activated to permit the generator output to stabilize under no load; this time delay is typically set for less than 1 min. Upon time-out, the load is transferred to the standby source.

Activation of another timer coincides with restoration of the normal source; this timer is typically adjustable from 0–30 min. The function of this time-delay relay is to provide an established minimum running time for the engine-generator any time that the unit is activated and to maintain the transfer switch in the standby mode, even if the normal source has been restored.

The purpose of this time delay is twofold. It is not uncommon for prolonged power outages to be preceded by a series of brief outages separated from each other by a matter of several seconds or a few minutes; the time delay prevents unnecessary cycling during such periods. It also ensures that the engine-generator will get a good workout under load, preserving it in good operating condition.

The transfer switch should be designed so that this feature is automatically nullified if the engine-generator set fails and power is available from the normal source.

A fourth timing function is recommended to permit the engine-generator to run unloaded for a while after transfer back to the normal source; an interval of 5 min or less is usually applied. This function permits the units to cool down properly and permits fan action to prevent excessive heat from building up in the generator. This feature is especially recommended for diesel-driven units.

If the standby system comprises more than one automatic transfer switch, it is desirable to sequence the transfer of loads from normal to standby source. This function, also performed by timers, can reduce the starting kVA requirements of the engine-generator set — an especially important consideration when the load is predominantly motors.

4.3.7 Voltage Tolerances. The basic standard for the voltage tolerance for utilization equipment is ANSI C84.1-1982 [1]. These limits are based on the T-frame motor with slightly reduced limits for equipment other than motors. In the case of special equipment, the manufacturer's tolerance limits should be obtained. Care should be taken in determining the minimum voltage for transfer switch operation to distinguish between equipment that is not damaged by low voltage, even though operation is unsatisfactory, such as incandescent lamps and resistance heaters, and equipment that may be damaged by low voltage or that may cause damage or unsafe conditions at low voltage.

4.3.8 Transferring Motor Loads. Transferring motor loads between two sources requires special consideration. Although the two sources may be synchronized, the motor will tend to slow down upon loss of power and during transfer, thus causing the residual voltage of the motor to be out of phase with the oncoming source. The speed of transfer, total inertia, and motor and system characteristics are involved. On transfer, the vector difference and resulting high abnormal inrush current could cause serious damage to the motor, and the excessive current drawn by the motor may trip the overcurrent protective device. Both motor loads with relatively low load inertia in relation to torque requirements, such as pumps and compressors, and large inertia loads, such as induced draft fans, etc, that keep turning near synchronous speed for a longer time after loss of power, are subject to the hazard of out-of-phase switching. Automatic transfer switches can be provided with various accessory controls to overcome this problem, including the following:

(1) Inphase transfer
(2) Motor load disconnect control circuit
(3) Transfer switch with a timed center off position
(4) Overlap transfer to momentarily parallel the power sources

Inphase transfer as shown in Fig 16 is commonly used for transferring low-slip motors driving high-inertia loads, provided that the transfer switch has a fast operating time. A primary advantage of inphase transfer is that it can permit the motor to continue to run with little disturbance to the electrical system and the process that is being controlled by the motor. Another advantage is that a standard double-throw transfer switch can be used with the simple addition of an inphase monitor. The monitor samples the relative phase angle that exists between the two sources between which the motor is transferred. When the two voltages are within the desired phase angle and approaching zero phase angle, the inphase monitor signals the transfer switch to operate and reconnection takes place within acceptable limits.

Motor load disconnect control circuits, such as shown in Fig 17 and similar relay schemes, are also a common means of transferring motor loads.

As Fig 17 indicates, the motor load disconnect control circuit is a pilot contact on the transfer switch that opens to de-energize the contactor coil circuit of the motor controller. After transfer, the transfer switch pilot contact closes to permit the motor controller to reclose. For these applications, the controller must reset automatically. The disconnect circuit should be arranged to open the pilot contact for approximately 0.5–3 s before transfer to the alternate power source is initiated.

Fig 16
Inphase Motor Load Transfer

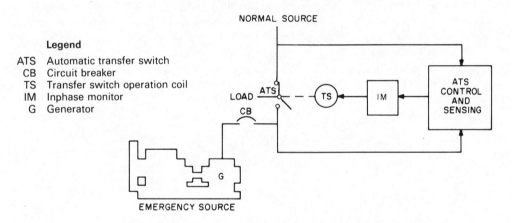

Fig 17
Motor Load Disconnect Circuit

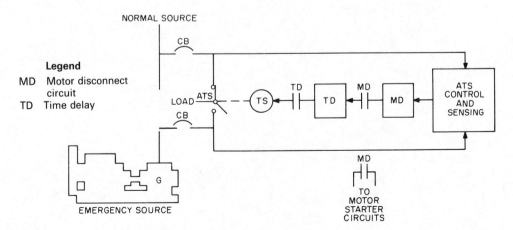

Transfer switches with timed center off (neutral) position are also used for switching motor loads; Fig 18 shows a typical arrangement. Interconnections between the transfer switch and motor controller are not required. Because there is no direct control of the motor controller, the motor controller may not drop out if it sees the residual voltage from the spinning motor. The off-time must be long enough to permit the residual voltage to reduce to a value (typically 25% of rated motor voltage) at which reconnection will not harm the motor, the driven load, or trip the breaker. ANSI/NEMA MG1-1978 [9] outlines a method of calculating safe values. The open circuit voltage decay times for a series of motors are shown in Fig 19. The center hold time should not be additive to the generator start-up time, so that the power outage is not needlessly lengthened by the center hold time of the transfer switch. Only when switching between two *live* power sources should the center hold function be in operation.

Overlap (closed transition) transfer with momentary paralleling of two power sources is shown in Fig 20. An uninterrupted load transfer provides the least amount of system and process disturbance. However, overlap can only be achieved when both power sources are present and properly synchronized by voltage, frequency, and phase angle. While the overlap arrangement is technically feasible, it is not always practical because of the complexity of the system and the cost of the relaying required.

Fig 18
Neutral Off Position

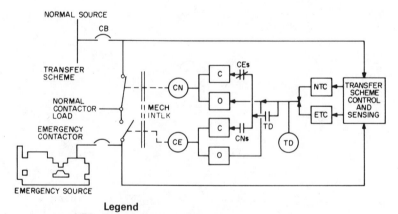

Legend

NTC   Transfer to normal control circuitry
ETC   Transfer to emergency control circuitry
CE   Emergency source contactor
CN   Normal source contactor
CEa   Emergency source contactor electrical interlock
CNa   Normal source contactor electrical interlock
TD   Time delay
C   Closing circuit
O   Opening circuit

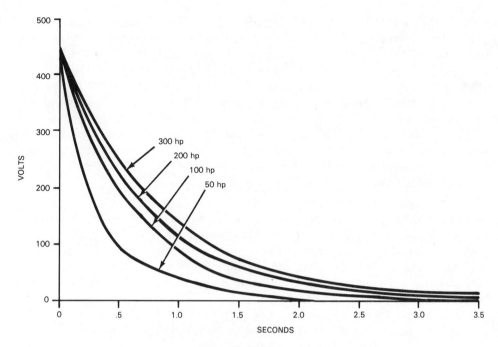

**Fig 19
Induction Motor Open Circuit Voltage Decay
(Based on Constant Speed)**

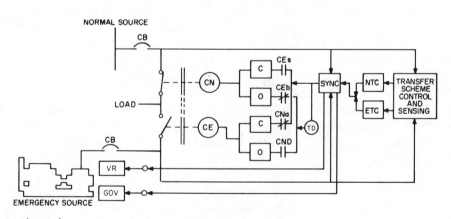

Legend

CEb   Emergency source contactor electrical interlock
CNb   Normal source contactor electrical interlock
  TD   Time delay to open alternate contactor if proper contactor does not operate to terminate parallel
        operation
SYNC   Automatic synchronizer to synchronize emergency source in voltage frequency and phase angle
        with normal source

**Fig 20
Closed Transition Transfer**

4.3.9 Multiple Utility Services. Figure 21 illustrates a typical system supplying electric power to a manufacturing plant. The system is designed for initial operation with utility line 1 utilized as the normal source and utility line 2 utilized as a normally open auxiliary source. The two utility lines are synchronized with each other so that they can be paralleled, but are not normally operated in this manner. However, unless the relaying is designed for it, the operation of the two incoming lines in parallel should be kept to a minimum — that is, the switching time.

Utility lines 1 and 2 enter the plant from a substation some distance away through underground conduits separated by about 1 ft and encased in the center of a 3×3 ft concrete enclosure 7 ft below the surface for protection.

Operation is as follows:

(1) If voltage on the normal source (line 1) drops appreciably for several cycles, the undervoltage relay will deactivate, trip the circuit breaker in line 1, and close the circuit breaker in line 2 (if acceptable voltage is present on line 2).

(2) When voltage is restored to line 1, the undervoltage relay is activated and initiates a timer. After the voltage has been present on line 1 for a predetermined time (usually 1–10 min), the circuit breaker in line 1 closes, after which the circuit breaker in line 2 opens.

(3) If there is no voltage present on line 2 when line 1 loses voltage, the circuit breaker in line 2 will not close. When voltage is restored to line 1, the circuit breaker in line 1 will immediately close.

(4) If a fault or overload occurs on the load side of either incoming circuit breaker, a lockout relay keeps both circuit breakers open, disabling the automatic transfer system until manually reset.

As power demands increase, this system can be expanded by inserting a tie circuit breaker in the 13.8 kV bus and additional relaying. Part of the load would then be supplied by each utility line with transfer of the entire load to the line feeder, should power be lost to one line. Load shedding of noncritical loads could be incorporated if necessary.

Operation would then be as follows:

(1) A loss of voltage or appreciable decrease of voltage on either utility line will cause that normally closed circuit breaker to open and the normally open tie circuit breaker to close. When normal voltage returns, the open utility line circuit breaker will close in a preset time (1–10 min), after which the tie circuit breaker will open.

(2) A simultaneous loss of voltage on both utility lines will cause both normally closed utility line circuit breakers to open and the normally open tie circuit breaker to close. A return to normal voltage on either utility line will cause that utility line circuit breaker to close in its preset time and the tie circuit breaker to remain closed. When normal voltage is established on the second utility line, that utility line circuit breaker will close in its preset time after which the tie circuit breaker will open.

(3) Fault current or overload current causing either utility line circuit breaker to open will also make the automatic closing feature of the tie circuit breaker inoperative until manually reset.

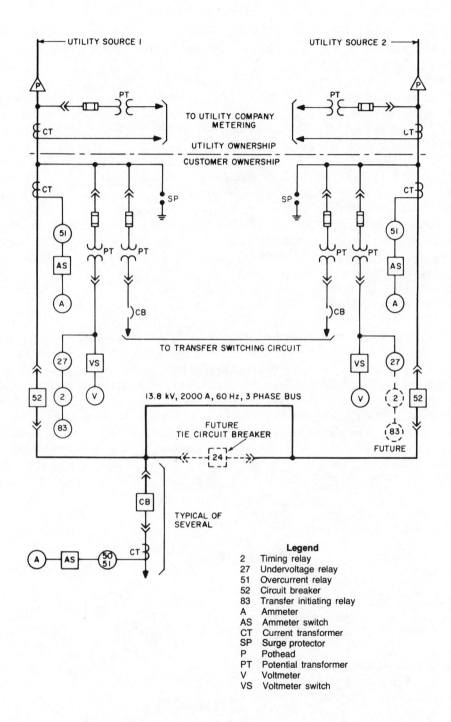

Fig 21
Typical System Supplying Electric Power to Manufacturing Plant

(4) If the utility company needs to take one line out of service, they notify the customer who then manually closes the tie circuit breaker and opens the line to be affected.

The arrangement shown in Fig 20 only provides protection against failure of the utility source. To provide protection against disruption of power within the building areas, it may be desirable to locate additional automatic transfer switches downstream and close to the load. A combination of circuit breakers and downstream automatic transfer switches is shown in Fig 14.

4.3.10 Bypass-Isolation Switches. In many installations, it is difficult to perform regular testing or detailed inspections on the emergency system because some or all of the loads connected to the system are vital to human life or are critical in the operation of continuous processes. De-energizing these loads for any length of time is difficult. This often results in a lack of maintenance. For such installations, a means can be provided to bypass the critical loads directly to a reliable source of power without downtime of the loads. The transfer switch can then be isolated for safe inspection and maintenance.

Two-way bypass-isolation switches are available to meet this need. These switches perform three functions:

(1) *Shunt the service around the transfer switch without interrupting power to the load.* When the bypass (upper) handle is moved to the bypass-to-normal (BP-NORM) position (Fig 22), the closed transfer switch contacts are shunted by the right-hand BP contacts. The flow of current then divides between the bypass and transfer contacts. This assures there will not be even a momentary interruption of power to the load should current no longer flow through the transfer switch, in which case the full current is immediately carried by the bypass contacts.

(2) *Allow the transfer switch to be electrically tested and operated without interrupting power to the load.* This can be done as shown in Fig 23. With the isolation (lower) handle moved to the test position, the load terminals of the

**Fig 22
Bypass to Normal**

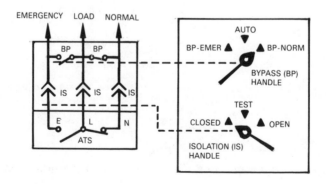

transfer switch are disconnected from the power source. The transfer switch is still energized from the normal and emergency sources and can be electrically tested without interrupting the load. The closed right-hand BP contacts carry the full load.

(3) *Electrically isolate the transfer switch from both sources of power and load conductors to permit inspections and maintenance of the transfer switch.* (See Fig 24.) With the isolation handle moved to the open position, the automatic transfer switch (ATS) is completely isolated. The load continues to be fed through the BP contact. With drawout capability, the transfer switch can now be completely removed without interrupting the load. In this mode, the bypass switch has a dual function. In addition to bypass, it also operates as a manual backup transfer switch.

Fig 23
Test Position

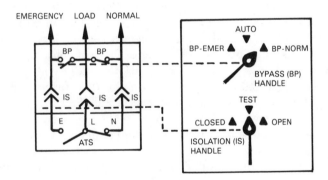

Fig 24
Complete Isolation

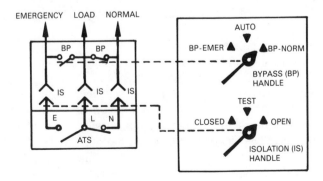

Although the foregoing illustrations show bypass to the normal source, the transfer switch can also be bypassed to the emergency source without interrupting power to the load, provided bypass is made to the source feeding the critical load. When bypassing to the emergency source, the same three functions can be performed after the transfer switch has transferred to emergency. While two-way bypass-isolation switch arrangements have been available for many years, only recently has it become possible to combine a two-way, noninterruption bypass function with the automatic transfer function all in one interconnected assembly.

The bypass and isolation portions of the switch assembly should incorporate *zero maintenance* design. This design concept avoids system shutdown during maintenance or repair. To achieve zero maintenance design, bypass contacts should be in the power circuit only during the actual bypass period. The objection to retaining the bypass contacts in the circuit at all times is that they, along with the bus bars, are also subject to damage from fault currents. While the transfer switch is repairable without disruption of service, the bypass switch is not.

Providing a combination automatic transfer and bypass-isolation switch in lieu of an automatic transfer switch more than doubles the cost.

4.3.11 Nonautomatic Transfer Switches. Nonautomatic transfer switches are used in applications where operating personnel are available and the load is such that immediate automatic restoration of power is not mandatory.

Some typical applications are found in health care facility equipment systems, industrial plants, sewage plants, civil defense control centers, farms, residences, communication facilities, and other installations where codes require approved devices. A typical installation of a nonautomatic transfer switch combined with several automatic transfer switches is shown in Fig 25.

Double-throw knife switches and safety switches are often not suitable for these applications. Many of these devices have limited capacity when switching between two unsynchronized power sources. Nonautomatic transfer switches generally have the same type of contacts and arc-quenching means as automatic transfer switches and meet ANSI/UL 1008-1983 [12].

Both electrically operated and nonelectrically operated switches are available. Electrically operated units are arranged for local or remote control station operation. Accessibility to the transfer switch is not necessary when it is remotely controlled by remote control stations. This may be an advantage in a large facility where the devices can be controlled from the plant engineer's operations room.

Two interlocked control relays are often included with the electrically operated switch to permit

(1) Line runs over small-gauge wire
(2) Low operating current through control station switch
(3) Partial voltage check before operating

Nonelectrically operated units are similar to the electrically operated type except for omission of electrical operation and inclusion of a quick-make/quick-break operator that can be manually operated from outside the enclosure. The speed of operation is similar to the electrically operated arrangement owing to the preloading of the main operating springs, thereby providing appropriate current make, break, and carry capabilities.

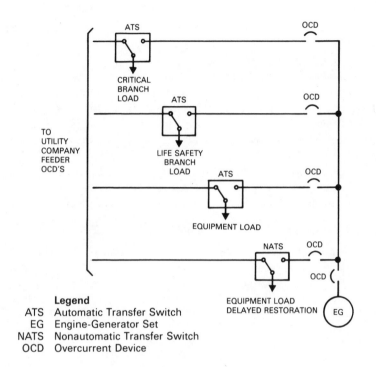

Legend
ATS   Automatic Transfer Switch
EG   Engine-Generator Set
NATS   Nonautomatic Transfer Switch
OCD   Overcurrent Device

**Fig 25
Typical Hospital Installation with a Nonautomatic
Transfer Switch and Several Automatic Transfer Switches**

4.3.12 Conclusion. Approximately 90% of transfer schemes designed to switch from the prime power source to the emergency source for commercial installations utilize conventional double-throw transfer switches. For maximum system reliability, the transfer switches are usually located close to the load rather than at the incoming prime power source. The use of interlocked service entrance circuit breakers for transfer schemes is usually limited to medium-voltage primary switching.

Two parallel inphase separate utility sources on line continuously with proper relaying on a normally closed tie circuit breaker provide increased reliability of electric power in some applications. This possibility should be investigated with the utility company representatives.

4.4 Turbine-Driven Generators

4.1.1 Introduction. Two general types of turbine prime movers for electrical generators are available, steam and gas/oil.

4.4.2 Steam Turbine Generators. Steam is usually not available if all electric power has been lost, although there are independent steam supply systems that may themselves have uninterruptible electrical systems. In this case steam might be considered. There are compact steam turbines that could bring power onto the line in about 5 min. This is a rather special source of supply, and details are not presented.

Steam turbines are used to drive generators that are larger than those that can be driven by diesel engines. However, steam turbines are designed for continuous operation and require a boiler with fuel supply and a source of water. Thus, they are expensive for use as an emergency or standby power supply and may have environmental problems involving fuel supply, noise, combustion product output, and heating of the condensing water.

Figure 26 shows an on-line steam turbine generator supplying a critical bus in parallel with one source of utility power. An alternate utility source can be manually switched on in a minute or so should there be a failure of the on-line utility source. The normal utility supply system should be large enough to supply the entire critical bus if the turbine is off. Relaying for such a system should be worked out with the utility. A contract is required with the utility to establish the right to parallel and the disposition of power purposely or inadvertently shipped to the utility. The system should function to interrupt generator input to utility faults, avoid out-of-synchronism, and detect islanding of the generator with a portion of the utility load.

Fig 26
Emergency and Standby Power System Using
Steam-Turbine and Dual-Utility Supply

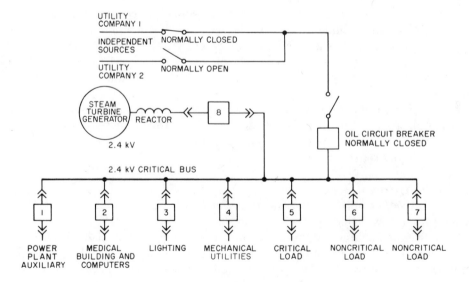

4.4.3 Gas and Oil Turbine Generators. The most common turbine-driven electric generator units employed today for emergency or standby power use gas or oil for fuel. Various grades of oil and both natural and propane gas may be used. Other less common sources of fuel are kerosene or gasoline. Service can be restored from about a 10 s minimum to several minutes, depending upon the turbine used.

Aircraft-type turbines driving generators have been commonly used where electric power may be needed for a few hours to days. Small industrial units have been developed. A small fuel storage facility for safety may be adequate, provided plans have been made by which additional fuel will be delivered when needed. Care should be taken to assure an adequate gas supply, should this be the source of energy, since an uninterruptible supply from a public utility company may be very expensive or unavailable. An interruptible supply may not be available when needed during cold weather. Earthquakes may destroy extensive underground utility distribution systems, but local storage may be intact. Environmental considerations may require the use of low-sulfur oil, which may be difficult to obtain.

Savings are possible by the installation of a turbine for emergency and standby power when used as a peak clipping unit to reduce the demand charge. This has the operating advantage of checking the unit often enough under load so that the operating personnel are familiar with the equipment and the unit is known to be ready for an emergency.

Turbine designs fall between two extreme categories. Aircraft-type engines embody highly sophisticated techniques for very light weight in relation to horsepower (0.25–0.50 lb/hp). This shortens life. Figure 27 shows a typical generator of this type with a riser diagram of typical loads served. Figure 28 shows a modular packaged gas turbine. The opposite philosophy employs the massive, bulky design techniques of steam turbines in an effort to assure long life, but with 10 lb/hp. Both have their place depending on the cost justification and hours of usage per year. Reliable industrial units are available within these ranges.

Accessory equipment for a gas turbine, such as filtering, silencing apparatus, and vibration mounts, may be required when climate conditions indicate contaminated dust-laden air or areas where noise and vibration level attenuation are required. Bands of noise from 75–9600 Hz are common and attenuation from 5–60 dB or more may be necessary in various bands.

A complete 750 kW gas turbine generator unit weighs approximately 13 000 lb and occupies less than 80 ft^2 of floor area. This factor of compact size and light weight can considerably reduce building costs and allow more efficient economic utilization of building space. Rooftop installations are both feasible and practicable.

Combustion turbines starting and loading are accomplished either automatically or manually. Warmup is not necessary.

There are four basic starting systems available for turbines:

(1) Electric motor supplied from batteries
(2) A small steam turbine
(3) A compressed air or gas system
(4) A small diesel engine

The controls for multiple-unit installations generally involve some interconnection considerations in a master control panel or synchronizing panel. The turbines

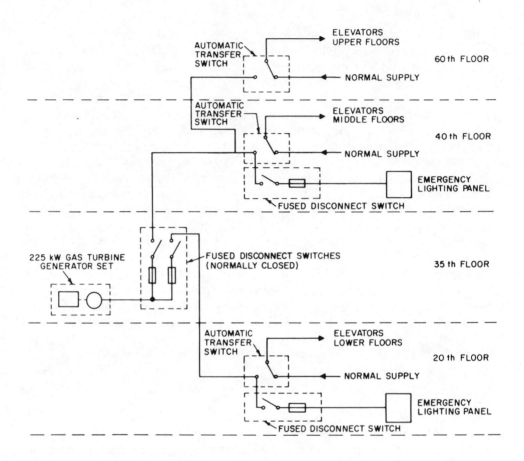

Fig 27
Typical Gas-Turbine-Generator and Riser Diagram of
Auxiliary Power System

may be programmed for automatic or manual operation, start sequence, and syn-chronizing, if desired.

Figure 29 shows the reduction in output capacity as gas- or oil-fired combustion turbines are installed at increasing altitudes. High air-input temperatures with the associated lower air density and low barometric pressures also reduce the available output. These limitations must be taken into account or the anticipated reliability and capacity may not be obtained.

Fuel consumption at sea level will be about 10 000–17 000 Btu/kWh output, depending upon size and variable factors. Thermal efficiency may be raised consid-erably if waste heat or recuperation can be used.

4.4.4 Advantages and Disadvantages. See 4.2.23 for merits of diesel engines versus gas-turbine prime movers.

**Fig 28
Modular Packaged Gas-Turbine-Generator Set Mounted on Trailer**

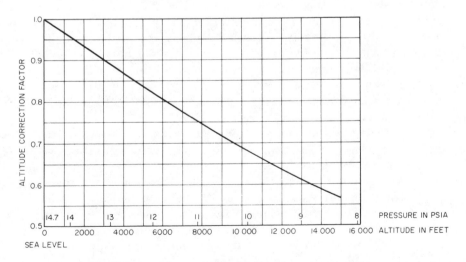

**Fig 29
Typical Performance Correction Factor for Altitude**

4.5 Mobile Equipment

4.5.1 Introduction. An important, and yet often overlooked, source of emergency or standby power is mobile equipment. For some industrial and commercial applications, it can be the simplest, most economical, and best solution for emergency or standby needs. Mobile equipment, in the broadest sense of emergency and standby equipment, could embrace portable transformers and even substations with switching and protective devices as are available and used by most utility companies. This discussion is limited to mobile equipment as a source for electrical power. Further, in recognition that some utility company generating equipment has been located on barges and transported on waterways as needed, this discussion is further confined to land-based mobile equipment. Widest usage, by far, for most industrial and commercial applications are engine-driven and turbine-driven generating equipment. Characteristics of both of these prime movers as stationary installations are discussed in detail in 4.2 and 4.4. It is the intention of this section to pinpoint the special requirements of those prime movers when they are adapted for mobility. Some special precautions that should be followed for mobile equipment are also outlined. The decision to have mobile equipment is generally based on having multiple usages. Thus, transportability to serve several load groups, with their varying needs and electrical characteristics, will place restraints on the design/selection process of such equipment.

4.5.2 Special Requirements. Mobile generating units can take almost any form from only a prime mover and generator assembly to a complete self-contained plant having its own transporting power. Smaller units can be considered mobile even if they are skid-mounted and have to be transported by loading on and unloading from another vehicle. Obviously, there is some limit beyond which it is not practical, either because of weight or physical size, to load and unload a unit for transporting to achieve the necessary mobility. It is also obvious that a point can be reached whereby generating equipment, because of its physical size and weight, should not be considered for mobile application, even if permanently installed on a truck-bed or a trailer assembly. In terms of capacity, the size break of practicality will vary with engine or turbine drives as prime movers, equipment handling facilities, and also with how much auxiliary equipment, for support of the unit, needs to be transported. Based on these premises, the following is an attempt to describe the several special requirements or needs of generating equipment when it is adapted for mobile applications.

Since mobile equipment may be required to be used outdoors or, when not in use, stored outdoors, it can be desirable to house it in a weather-proof enclosure. If its use may be in a sound-sensitive area, provisions for noise attenuation may be necessary. Whatever type of enclosure is employed, special precautions may be necessary to assure that the unit will have sufficient combustion and cooling air to operate within its rating. It may be necessary to equip the unit with louvers that are arranged to open automatically when the unit is running. For units where sound-proofing is required, a well-designed forced-air ventilating system will probably be needed. For application in sound-critical areas, such as residential, an appropriate high sound attenuating exhaust silencer or muffler may be warranted.

To avoid possible duplication of fuel-storage facilities in all locations where the unit will find use, it may be desirable to provide fuel storage with the unit. Storage capacity will depend on each need; however, an 8 h supply as a minimum would seem to be a reasonable period within which a portable fuel transport vehicle could be dispatched to replenish the fuel supply. Fuel oil for a diesel engine is a good fuel choice because of its availability, transportability, and storability during periods of disuse and because it presents less hazard than gasoline or natural and bottled gases. Tank capacity should be based on the unit's consumption when operating at full load.

If the mobile unit is to be a towed vehicle, it will probably be required to meet all of the state's (and possibly Interstate Commerce Commission's) requirements for such items as lights and markings. The vehicle should be designed for a realistic road speed, have stop lights, and probably a braking system. Thus, the plug receptacle wiring connections and the hitch assembly need to be compatible with similar mating equipment of the towing vehicle. Figure 30 shows a typical trailer-mounted unit in the range of 15–45 kW.

Figure 31 shows a 2800 kW turbine-generator unit with self-contained switchgear.

A careful selection of the generator output voltage, or voltages, phases, and frequency should be made to be compatible with all envisioned load requirements. Multiwinding generators can be specified to enable field connection and selection of voltages. Most mobile rental units have multiple windings and are reconnectable for common utilization voltages.

Often with smaller mobile units it is advantageous to transport the cable connection with the generating equipment. If connection time is critical in restoring electrical service, the cable connection can be hard-wired at the generator terminals, transported on an integral reel, and equipped with a plug to mate with the load-end receptacle. That receptacle could, in turn, be hard-wired into the load facility's electrical system. Many telephone building facilities and switching centers are arranged for such a plug-in connection of a mobile standby unit. The cable length

Fig 30
Typical Trailer-Mounted Model
(15–45 kW Capacity)

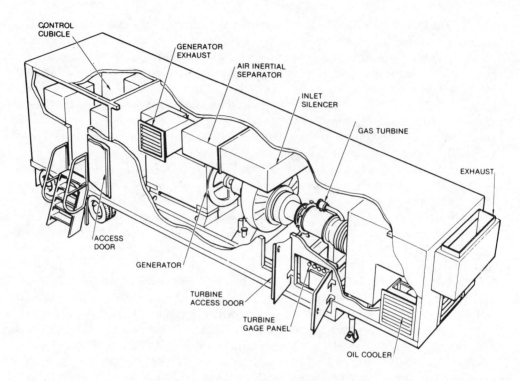

Fig 31
Typical 2800 kW Mobile Turbine-Driven Generator Set

should be determined for the extreme location where the equipment will be used. If a cable reel is used, the diameter should be selected after consideration of the manufacturer's recommended minimum bending radius for the cable. The cable should be of a portable variety as recognized by the NEC [10].

A disconnect switch or a circuit breaker is often installed at the generator. See Chapter 6 for information on generator protection.

Speed control through governor and fuel supply systems selection for prime movers can be a critical choice for a mobile unit if no consideration of the demands of the connected load is made. The same is true in a choice of voltage-regulating methods in combination with excitation systems for the generator. Suffice it to say that a more extensive analysis, before selection, should be made for a mobile unit in comparison with a stationary installation. This is because of the several different possible demands that may be placed on a mobile unit through the more diversified demands of the loads. Improper selection of speed control or voltage control, or both, could mean either money spent wastefully or could cause the unit to be unresponsive to the more critical demands of some loads.

It is generally advantageous to arrange mobile units for starting with self-contained electric storage batteries because of their transportability. A compressed air start unit would require a fairly large volume air receiver that would usually reach such large size and high weight proportions that its transportability would be difficult. If batteries are used, the permanent storage location, used when the mobile unit is not in service, should be arranged with a battery-charging unit.

If a reciprocating engine unit is to be used or stored outdoors, an ethylene glycol solution and an immersion heater should be used in the cooling system.

4.5.3 Special Precautions. Mobile equipment, when compared with stationary equipment, will have some precautions that should be planned for and followed. Many of the items listed below are somewhat obvious but nevertheless often overlooked, especially in the early stages of planning for the mobile unit's utilization and its related storage area.

If a single area is used for storing the unit when not in service, it should be equipped with a means of heating the coolant and charging the unit's batteries through a battery charger and a flexible plug and receptacle connection. If an air start is planned, an air compressor with flexible high-pressure lines and connectors should be used. If the storage space is unheated and outdoors, in colder climates it may be advantageous to provide branch circuit wiring for connection of a unit's immersion-type heater located in the unit's cooling water jacket. If there are extended periods when the unit is not used, it may be advantageous to have a dummy load bank for maintenance testing of the unit under load. Other additional features that could prove valuable for installation at the storage facility are an annunciator unit, which monitors generator set key functions, and an automatic exercising unit.

If a plug-in unit is to be planned, it is important that each load facility, where a mobile unit is to be used, be wired with a receptacle that mates with, and is rated for, the plug on the unit. It should be connected to the load through a transfer switch. Because of inherent characteristics of a mobile generating unit, when compared with a facility's normal source (often a utility company), staged sequencing of returning the facility's loads to service may be in order. Rarely will automatic load sequencing be preferred over manual sequencing through a well-planned start-up procedure.

Assuming that, in initial planning, it may be possible at the load facility to anticipate the operating location of mobile equipment, the fresh air intakes for an adjacent or adjunct building should be physically well-separated from the operating location for the unit.

4.5.4 Maintenance. Since mobile equipment's usage is often unscheduled, necessary maintenance and testing procedures can become sloppy with important items neglected. The unit should be kept ready to go at all times. If the unit is for backup of life safety equipment, it should be treated as similar, stationary, permanently installed equipment employed for the same function. As such, a regularly scheduled operating run of the unit for some representative duration under load will enable checking of such important functions as prime mover lubrication and cooling, and charged batteries or starting systems, or both. This testing can often be accomplished through automatic exercising and alarm annunciation equipment installed at the storage location.

4.5.5 Application. Mobile equipment finds widespread application on multi-faceted and physically scattered facilities such as might be operated by a municipality. Many municipalities will have such diversified operations, all of which will, on occasion, require emergency and standby power, such as sewage and waste water treatment and pumping plants, garbage and refuse collection and disposal, central fire-alarm stations, centralized traffic-control facilities, convention, sports, and recreational complexes, and power and steam generating plants. Other users of mobile equipment might include college and educational facilities, military bases, and large governmental, institutional, and industrial complexes, all typified by scattered individual buildings, most having their own electrical service and, in many cases, privately owned electrical distribution facilities between buildings.

4.5.6 Rental. Often the expenses of purchase of a mobile unit cannot be justified when measured against how often the equipment is needed and used. There are several sources for the rental of mobile generating equipment, namely, equipment manufacturers' distributors and service organizations, some utility companies, and large construction contracting firms. Rental houses generally charge approximately 10% of the list price of the equipment per month based on an 8-h day. The weekly rate is generally ⅓ the monthly rate and the daily rate ⅓ the weekly rate. For more than 8 h per day, the rate is 1½–2 times the 8-h rate.

Sometimes power is lost from the normal utility supply source, but a lower or higher voltage is available close by. A mobile transformer can be used for emergency power secured in time to prevent damage from freezing or in a quantity sufficient for partial production.

4.5.7 Fuel Systems. Growing shortages of gas and liquid fuel sources require advanced planning for a firm. A dependable source of fuel supply, readily available under any foreseeable emergency condition, should exist. When natural gas is planned, a firm source should be available; a back-up source of propane should be considered. Where liquid fuels are planned, readily available storage at the point of usage should be considered. Since liquid fuels deteriorate during long storage, a system of use and replacement to assure a fresh supply should be planned.

4.5.8 Agricultural Applications. Farm standby tractor-driven generators are available, practical, and usually reasonable in cost, since the prime mover is serving a dual purpose and is usually on hand for other uses.

4.6 References. This standard shall be used in conjunction with the following publications:

[1] ANSI C84.1-1982, American National Standard Voltage Ratings for Electrical Power Systems and Equipment (60 Hz).

[2] ANSI/IEEE C37.95-1973, IEEE Guide for Protective Relaying of Utility–Consumer Interconnections.

[3] ANSI/IEEE Std 100-1984, IEEE Standard Dictionary of Electrical and Electronics Terms.

[4] ANSI/IEEE Std 141-1986, IEEE Recommended Practice for Electric Power Distribution for Industrial Plants.

[5] ANSI/IEEE Std 142-1982, IEEE Recommended Practice for Grounding of Industrial and Commercial Power Systems.

[6] ANSI/IEEE Std 241-1983, IEEE Recommended Practice for Electric Power Systems in Commercial Buildings.

[7] ANSI/IEEE Std 387-1984, IEEE Standard Criteria for Diesel-Generator Units Applied as Standby Power Supplies for Nuclear Power Generating Stations.

[8] ANSI/NEMA ICS2-1983, Industrial Control Devices, Controllers, and Assemblies.[13]

[9] ANSI/NEMA MG1-1978, Motors and Generators.

[10] ANSI/NFPA 70-1987, National Electrical Code.

[11] ANSI/NFPA 101-1985, Life Safety Code.

[12] ANSI/UL 1008-1983, Safety Standard for Automatic Transfer Switches.

[13] EGSA 101E-1984, Glossary of Terms — Electrical.

[14] EGSA 101M-1984, Glossary of Terms — Mechanical.

[15] EGSA 101S-1982, Standard Specifications for Standby Engine Driven Generator Sets.

[16] EGSA 101P-1985, Engine Driven Generator Set Performance Standard.

[17] HEISING, C. R. and JOHNSTON, Jr, J. F. Reliability Considerations in Systems Applications of Uninterruptible Power Supplies. IEEE Transactions on Industry Applications, vol IA-8, Mar/Apr 1972, pp 104–107.

[18] IEEE Committee Report. Report on Reliability Survey of Industrial Plants, Part I: Reliability of Electrical Equipment, *IEEE Transactions on Industry Applications*, vol IA-10, Mar/Apr 1974, pp 213–235.

[19] SAWYER, J. W. Gas Turbine Emergency/Standby Power Plants. *Gas Turbine International*, Jan/Feb 1972.

4.7 Bibliography

[B1] DAUGHERTY, R. H. Analysis of Transient Electrical Torques and Shaft Torques in Induction Motors as a Result of Power Supply Disturbances. *IEEE Transactions on Power Apparatus and Systems*, vol PAS-101, no 8, Aug 1982.

[B2] *Engineer's Guidebook to Power Systems*, Kohler Company, Kohler, WI.

---

[13] ANSI/NEMA publications can be obtained from the Sales Department, American National Standards Institute, 1430 Broadway, New York, NY 10018, or from the National Electrical Manufacturers Association, 2101 L Street, NW, Washington, DC 20037.

[B3] GILL, J. D. Transfer of Motor Loads Between Out of Phase Sources. *IEEE Transactions on Industry Applications*, vol IA-15, no 4, July/Aug 1979, pp 376–381.

[B4] JOHNSON, G. S. Rethinking Motor Transfer. Conference Record IEEE Industry Applications Society, Oct 1985, Toronto, Ontario.

[B5] McFADDEN, R. H. Re-energizing Large Motors After Brief Interruptions — Problems and Solutions. 1976 Pulp and Paper Conference, San Francisco, CA.

[B6] Onan Power Systems Manual, Onan Corporation, 1400 73rd Avenue, NE, Minneapolis, MN.

Chapter 5
Stored Energy Systems

5.1 Introduction. Energy usable for electric power generation may be stored in many ways. Liquid and gaseous fuels are a form of energy stored for use in engines and turbines to turn generators. Steam and high-pressure gases may be stored for use in place of utility power. For the purpose of this chapter, we will study the two most prevalent stored energy systems, mechanical stored energy systems and battery systems.

Mechanical systems store energy in the form of kinetic energy in a rotating mass. This energy is converted to useful power by using kinetic energy to turn an ac generator.

Battery systems store energy in the form of electric potential in a bank of battery cells. This energy may be converted to useful power by one of two means, either by a dc motor rotating an ac generator, or by an electronic dc–ac converter, called a static inverter. This chapter also discusses methods by which motor-generator sets and static inverters have been configured into power conditioning systems and uninterruptible power supplies (UPSs). Each system is designed to meet one or more of the following goals:

(1) To filter, regulate, and condition power for sensitive computers and other electronic equipment

(2) To isolate the load from the power source

(3) To permit orderly, controlled shutdown of equipment in the event of a power failure

(4) To bridge the interval from the occurrence of a power failure until an emergency standby generator can start and assume load

(5) To provide continuous power uninterrupted by power failures

Note that power conditioning and isolation are important benefits of energy storage systems. In fact, more UPS systems are installed for *power conditioning purposes* than for backup power purposes since 99% of all power problems do not require the backup time of a battery bank or diesel generator. So, in selecting a system, careful attention should be paid to the quality of output power in terms of harmonic distortion, stability, overload capacity, overload protection, and reliability.

139

The primary difference in the usability of the two energy storage systems is that mechanical stored energy has a practical time limit of less than 10 seconds, while battery banks can be paralleled to achieve time limits of many minutes or many hours.

For this reason, *most* uninterruptible power supply systems (whether rotating or static) use batteries as their primary backup power source. One notable exception is the engine-generator/motor-generator system described in 5.5.9. The other exception is the dual-feed UPS described in 5.5.10. All other systems available today convert dc battery power to ac power through an inverter or an ac generator, or some combination of the two.

This chapter addresses these systems first, by discussing the two energy storage systems, inertia and battery. Then, by discussing the two basic methods available for conversion of that stored energy to ac power, ac generator or static inverter. Static uninterruptible power supplies (UPSs) are discussed in 5.4. Motor-generators and rotating UPS systems are discussed in 5.5. Both types of UPS systems accomplish the goals listed above.

5.2 Mechanical Energy Storage

5.2.1 Introduction. Significant amounts of energy may be stored in the form of kinetic energy (inertia) in a rotating mass. This energy is created by an ac or dc motor imparting rotational energy to its rotor and the rotor of an ac generator and a flywheel, if one is used. This stored kinetic energy can be converted from inertia to useful ac power by the rotating ac generator.

5.2.2 Kinetic Energy. The energy contained in a rotating mass is defined by the equation:

$$\text{energy} = 1/2(JA^2)$$

where

J = the moment of inertia, in kilogram-meters^2
A = the angular velocity, in radians/second.

This energy is created by an ac or dc motor turning itself and the ac generator. The rotors and shaft(s) of the motor and generator make up the rotating mass. A flywheel may be added to increase the mass. The motor and generator may be coupled in a variety of ways. Couplings such as pulleys and belts increase the rotating mass.

The resulting stored kinetic energy (inertia) will continue turning the ac generator in the absence of power to the motor. The length of time that useful power is generated (called ride-through time) is determined by some key criteria:

(1) The inertia of the rotating mass
(2) The percentage of full-rated load placed on the ac generator
(3) The acceptable tolerance on the output frequency

For motor-generators under full load with a tolerance of –1 Hz maximum underfrequency, and no flywheel, typical ride-through times range from 50–100 ms. The addition of a flywheel can raise the time to 0.5 seconds or more. Flywheel motor-generators have been built with ride-through times of many minutes, using a constant r/min coupling. They have not proven commercially useful.

5.3 Battery Systems

5.3.1 Introduction. A battery is the most dependable source available for emergency or standby power and, when applied with other devices, can also be one of the most versatile.

There are various categories of batteries, that is, automobile, aircraft, marine, motive power industrial, and stationary industrial. The stationary industrial battery is primarily used to supply backup power for communications, emergency lighting, fire alarms, switchgear, generator set cranking, instrumentation, programmable logic controllers, distributive control systems, etc.

A stationary battery system is comprised of three interconnected and interdependent subsystems: battery, charger, and load.

5.3.2 Stationary Battery Construction. Stationary batteries are configured by series connecting a group of individual cells to make up the desired voltage.

There are basically two different types of batteries used in stationary industrial applications: one is the lead-acid electrochemical couple, and the other is the nickel-cadmium electrochemical couple. There are, of course, variations of each type and in turn various advantages and disadvantages for the different battery types. For a detailed discussion of these factors, the local battery sales engineer should be contacted.

It can be briefly stated that the lead-acid battery is less expensive to purchase than its nickel-cadmium counterpart. However, this initial capital cost may be offset in many applications because nickel-cadmium batteries generally exhibit a longer life, more rugged construction, and lower maintenance. The lower maintenance cost of nickel-cadmium batteries may be debatable due to the requirement of needing additional batteries to obtain the necessary voltage.

Three basic designs are employed in stationary lead-acid battery positive plate construction: Fauré, Planté, and multitubular. The negative plate of the battery undergoes relatively little change, and virtually all manufacturers have standardized on Fauré construction for negative plates.

Fauré plates are made in two versions, lead-antimony and lead-calcium. In either version, an alloy of lead and antimony or calcium is pasted onto a flat lead grid. The advantage of the lead-antimony is its ability to support prolonged and frequent discharges with a minimum of structural change; it has the disadvantage of requiring frequent addition of water as the cell ages. The lead-cadmium version requires very little watering throughout its lifetime, but frequent prolonged discharges can create structural growth that can shorten the battery's life.

Stationary nickel-cadmium batteries are normally constructed in a pocket plate design. Other designs, such as the sintered plate design may have an undesirable feature referred to as a *memory*. This effect is described later.

The lead-acid electrochemical couple is nominally a 2 V couple. The nickel-cadmium battery is nominally a 1.2 V couple. Therefore, more nickel-cadmium cells would be utilized to configure a given battery than would be required for a lead-acid battery.

The number of cells in a battery for any specific system is a matter of adapting to suit the voltage available for charging and the voltage required at the end of the discharge period (voltage window). The most frequently used systems encountered and the number of cells normally applied are given in Table 13.

5.3.3 Recharge/Equalize Charging. In lead-acid batteries, even if the battery is not discharged, the individual cell voltages will begin to drift apart and approximately every 60–90 days the lower voltage cells will need to be brought back to full charge by increasing the charger voltage approximately 10% for 25–30 hours. This is referred to as *equalizing* the battery. Nickel-cadmium batteries have much less self-discharge, and as a result, if the nickel-cadmium battery is not discharged with an external load, it will remain fully charged for many years at 1.4 V/cell. Therefore, nickel-cadmium cells do not need to be equalized.

However, it must be understood that nickel-cadmium batteries do need the dual-rate charging mode of the *float/equalize* battery charger.

Whether lead-acid or nickel-cadmium, both batteries need approximately 10% higher voltage to restore the discharged battery to a fully charged state.

Standby batteries are normally applied in *float* operation; the battery, battery charger, and load are connected in parallel. See Fig 32. The charging equipment is sized to provide all of the power normally required by relatively steady loads, such

Table 13
Number of Cells for Desired Voltage

Nominal battery voltage	120	48	32	24	12
Number of lead-acid cells	60	24	16	12	6
Number of nickel-cadmium cells	92	37	24	19	10
Equalize/recharge voltage	143	58	38	30	15.5
Float voltage	129	51	34	26	13
End voltage*	105	42	27	21	10.5
Voltage window	143–105	58–42	38–27	30–21	15.5–10.5

*The end voltage is a limit imposed by the manufacturer of the electrical equipment being powered. However, as a general rule of thumb, lead-acid cells should not be discharged below 75% of their nominal voltage (1.5 V per cell), and the pocket plate nickel-cadmium cells should not be discharged below 50% of their nominal voltage (0.6 V per cell). To avoid deep discharge, most battery systems include an undervoltage relay that automatically interrupts discharge at a preset end voltage.

NOTE: It is not uncommon to vary the number of cells for a specific application.

Fig 32
Battery "Float" Diagram

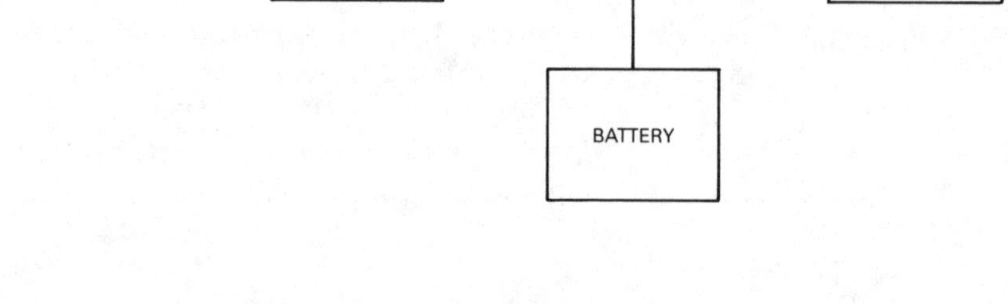

as indicating lamps and relay holding coils, and minor intermittent loads, plus enough additional power to keep the battery at full charge. Relatively large intermittent loads will draw power from the battery; this power will be restored to the battery by the charger when the intermittent load ceases.

When ac input power to the system is lost, the battery instantly assumes all of the connected load. If the battery and charger are properly matched to the load and to each other, there will be no discernible voltage dip when the system reverts to full battery operation.

When an emergency load on the system ends and charging power is restored, the charger delivers more current than it would if the battery were fully charged. The charger must be properly sized to ensure that it can serve the load and restore the battery to full charge within an acceptable time. The increased current delivered by the charger during this restoration period will taper off as the battery approaches full charge. The charger controls will maintain the battery float voltage at the prescribed value when the battery is fully charged.

A single-rate float charger will adequately maintain a fully charged nickel-cadmium battery until it is discharged by an external load. However, once the battery is discharged, it will not recharge to more than about 85% at float voltage regardless of the current capacity of the charger. With each successive discharge, the nickel-cadmium battery in such a charging circuit may continue to lose capacity. This phenomenon has from time to time been referred to as *memory effect*. It is simply a result of inadequately recharging any battery. It is even experienced in lead-acid batteries. However, usually before the loss of capacity is noted, the lead-acid battery is destroyed by sulfation of the positive plates, which is a rapid result in an undercharged lead-acid battery.

The charging rectifier, or battery charger, is a very important part of the emergency power system, and consideration should be given to redundant chargers on critical systems. Refer to Fig 33 for a typical redundant system. A general formula for sizing the battery charger for an inverter system would be as follows:

battery charger (amperes)

$$= \frac{\text{inverter output (VA)} \times 100}{\text{voltage input} \times \text{conversion efficiency}} + \frac{1.15 \times \text{battery capacity (A} \times \text{h)}}{\text{desired recharge time (h)}}$$

The battery charger output should be derated for both altitude and temperature. These requirements have to be recognized since the user frequently establishes these conditions. A larger than normal rating may be required to make up for the reduced capacity. A typical derating graph is shown in Fig 34.

5.3.4 Battery Sizing. To properly size any battery, the duty cycle should be defined with respect to the following:

(1) How many amperes?
(2) For how long?
(3) To what end voltage?
(4) At what temperature?

The size of the battery required depends not only on the size and duration of each load, but also on the sequence in which the loads occur.

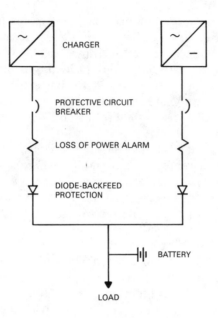

Fig 33
Typical Redundant Charger Circuit

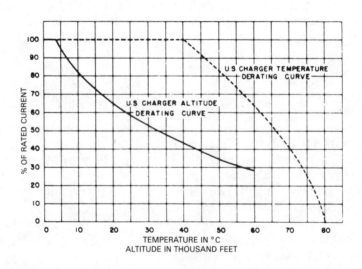

Fig 34
Derating Curves for Battery Chargers Due to Altitude and Temperature

The battery is sized to support the critical load until either the critical load can be shut down in an orderly manner, or the utility power returns, or an alternate standby source can be started and connected. Typical battery support times might be 5, 10, or 30 minutes. Rather than purchasing larger battery capacity, an engine or turbine generator standby power might be considered.

The battery system should be sized from manufacturers' data for a particular application in a known operating temperature range. Most ampere-hour ratings are for a temperature of 77 °F and a reduction of ampere-hour capacity is usually necessary for operation at lower temperatures. Some manufacturers derate their lead-acid batteries by as much as 60% from the 77 °F ratings for operation at 0 °F.

Ampere-hour (AH) capacities decrease as the rate of discharge increases. Therefore, a simple summation of the various loads (area under the current/time curve) may yield an undersized battery.

For varying loads, a summation of the various loads should be made as follows:

$$AH = A_1 T_1 + A_2 T_2 + \cdots + A_n T_n$$

where

    AH = ampere-hours
    A = load, in amperes
    T = time, in hours

A larger capacity battery will be required if there is a large discharge rate at or near the end of the cycle. Therefore, to verify that the battery selected is adequate, it should be checked by starting at the beginning of the discharge cycle and subtracting the energy (AT) removed by each load in order to determine if adequate capacity still remains for the final load interval.

Experience has shown that the lead-calcium battery requires several days to weeks to return to full equal charge on all cells following a discharge to the rated end voltage.

Other battery types may be required where frequent prolonged power outages occur, since there may not be time to fully recharge between outages without raising the charging voltage beyond the rating of the connected load.

Lead Planté and lead tubular batteries are also available, but their use is limited. Thus they are not included in Table 14.

ANSI/IEEE 450-1987 [5][14] and IEEE Std 485-1983 [6] provide additional guidance on age compensation and sizing, respectively.

5.3.5 Unit Lighting Equipment. For use in small- to medium-sized installations, hallways, stairwells, and equipment rooms, inexpensive self-contained battery units are available. These units are available in a variety of configurations and styles, from the standard metal cabinet with attached lighting fixtures, as shown in Fig 35, to more decorative styles.

---

[14] The numbers in brackets correspond to those of the references listed at the end of this chapter; when preceded by B, they correspond to the bibliography at the end of this chapter.

Table 14
General Differences for Various Battery Types

Battery Type	Physical	Typical Characteristics
Lead–calcium	Pasted lead-calcium Positive plate Sulfuric acid Electrolyte	Life 12–15 years, poor in high temperatures or many or deep discharges. Lowest water loss of any lead battery. Lowest cost.
Lead–antimony	Pasted lead-antimony Positive plate Sulfuric acid electrolyte	Life 10–12 years, good for cycle applications. Medium cost.
Nickel–cadmium	Pocket plate construction Nickel positive plate Cadmium negative plate Potassium hydroxide Electrolyte	Life 20–23 years, good for high or low temperatures. Superior for short fast discharges. Superior for deep discharges or for many cycles. Can be rapidly recharged. Highest cost.

NOTE: The end of battery life is defined as follows: When a rechargeable cell has been fully recharged and discharged in a load test and it fails to provide 80% of its original rated capacity, it has failed.

Fig 35
Typical Battery Unit

A variety of battery types are available for use in these units, such as wet lead–antimony, lead–calcium, and pocket plate nickel–cadmium. All of these require limited maintenance with the addition of water periodically. Because scheduled maintenance is difficult to ensure, the trend today is to utilize the maintenance-free type batteries, such as sealed lead–calcium, pure lead, and sealed nickel–cadmium.

These units are generally 6 V dc with a limited number available in 12 V dc. All of these units consist of a small automatic charger to maintain a proper charge of the battery. In the event of a power failure, the lamps are automatically switched on and supplied from the battery. When normal power is restored, the lamps are turned off and the battery is automatically recharged. Performance requirements are given in ANSI/NFPA 70-1987 [7] (National Electrical Code [NEC]).

Some units have provisions of powering remotely mounted lamps and exit signs, in addition to the lamps mounted on the unit itself.

5.3.6 Central Battery Lighting Systems. Building-wide systems may also be used to connect many lamps to a central battery charger console. The decision may be to install many 6 V units or a single 12, 32, or 115 V system with a power source (battery, charger, console) centrally located. The trend today, particularly in new construction where 10–100 lamps are involved, is to utilize the central system. The advantages of such a system are as follows:

(1) Centralized power source, eliminating the need for single units located throughout the building, use of less space (only lamps are in the areas to be protected), and remote location of the power source. This facilitates maintenance and testing of the system.

(2) Availability of alarm and protection circuits, which increases the flexibility of the system.

(3) Advantage of distributing at lower current, higher voltage, resulting in decreased losses.

(4) Central location, which reduces maintenance costs and provides a means for regular testing of the system for reliability.

When systems larger than the 32 or 115 V are justified, the inverter system is recommended.

5.3.7 Factors to Consider When Selecting Emergency Lighting Systems. Examples of immediate questions are what area is to be protected and what light level is necessary. The layout of the building, department, and usage of rooms will answer these questions. Areas that should be lighted are places near moving machinery, passageways, exits, stairwells, etc. The light level required is dictated by several factors, including applicable codes, usage, available funds, and personal preference. The minimum light level, where required by ANSI/NFPA 101-1985 [8], is 1.0 fc. The light level is determined by the number of lamps, their wattage, the lumen output of the lamps/watt, the reflector efficiency, the reflectance of the building surfaces, and the area covered.

As the number of lamps, length of wire runs, and the size of lamps increase, the voltage of the system should be increased to obtain a minimum cost and a more efficient system. The 32 V system should be applied on intermediate size applications; larger applications may require the 115 V system or inverter system.

In summary, there are three steps in sizing an emergency lighting system:

(1) Determine number, location, and type of lamps needed

(2) Determine total lamp wattage of all selected lamps

(3) Determine protection time required

Larger systems require a distribution panel. Its purpose is to distribute the available dc power to a number of individually fused circuits, with each fuse preferably monitored by a visual and audible alarm circuit. If one circuit suffers a fault, the fuse protection removes the faulted circuit from the dc bus, allowing the battery to power the remaining circuits. The distribution panel is usually located in close proximity to the battery supply.

The normal ac voltage supply to the battery charger and the battery voltage should be continually monitored. An alarm should be sounded whenever the supply voltage fails and if the battery voltage should become high or low. Permanent battery damage can result from continued operation with the plate surfaces exposed to the air. Therefore, an alarm system to monitor electrolyte level in the battery should be considered. The battery system alarm may be in any convenient location.

5.3.8 Multiple Sources Used for Normal Lighting. The systems described up to now are used whenever a single voltage source is used for normal lighting. If the load contactor senses the failure of the ac voltage it monitors, all emergency lamps will come on.

There are systems that will monitor each line-to-neutral voltage of a three-phase system. Thus, upon failure of any one or all three of the voltages, the load contactor will close, turning on all emergency lights.

If it is not desirable to turn on all the emergency lights, but only those of the room or zone where the normal lighting has failed, a multiload contactor panel may be used to monitor the ac voltage that supplies the lighting in the area or zone protected by that load contactor. If that voltage should fail, then that contactor only will close, and its lights will come on. If more than one ac voltage fails, then more than one string of emergency lights will come on.

When the normal lighting is supplied by mercury vapor or other high-intensity discharge (HID) lamps, full light will not be available during the ignition period. This period could range from 1–20 min, depending on temperature, line voltage, and the type of lamp. The normal characteristic of an emergency lighting system is to supply lighting only when the normal power is not available; as soon as it returns, the emergency lamps will go out. Thus there will be a period of time that no lighting will be available. To overcome this defect, a time delay on the load contactor should be considered.

The battery will try to continue to support power as long as the normal power is not available. In actual practice, however, the battery is limited in the amount of power it can deliver. A power failure of long duration will result in an overdischarged battery. Nickel–cadmium batteries are not affected by this overdischarge, but a lead–acid battery may have permanent damage as a result of this overdischarge. In this case, lead–acid batteries should be protected by utilizing a dropout relay. This relay is sensitive to voltage and disconnects the battery once its voltage reaches a critical point, generally 87.5% of system voltage.

5.4 Battery/Inverter Systems

5.4.1 Introduction. Continuous and disturbance-free ac power is required for a growing number of critical load applications. The foremost example of this need is sensitive electronic equipment used for data processing, life support, communication, and control functions (see 3.10–3.13).

Combining the energy storage capability of batteries with solid-state inverter technology provides a reliable, high-quality ac power system. Because of the capability to operate continuously on-line, affording no break in power when the primary source fails, these systems have been designated as static uninterruptible power supplies (UPS).

5.4.2 Battery/Inverter Supply Used as Standby Source. Figure 36 shows a relay transfer system for supplying an emergency power source to a load upon the loss of the prime source. Power to the load will be interrupted, as shown in Fig 37, from 60–190 ms, depending upon the type and size of the dc contactor and transfer switch. More costly but faster transfer systems use static switches.

In the system of Fig 36, the loss of voltage on the ac line will cause the dc contactor to close and supply power to the inverter. At the same time, the transfer switch operates to transfer the load to the inverter supply. This system is adequate for lighting, signal circuits, radio systems, and other loads that can tolerate the short interruption of power.

5.4.3 Nonredundant Uninterruptible Power Supply. The basic UPS configuration consists of a single rectifier, battery, and inverter operating continuously in the powerline. These are available in sizes ranging from 250 VA to over 500 kVA. A typical one-line diagram is shown in Fig 38.

Fig 36
Short-Interruption Standby System

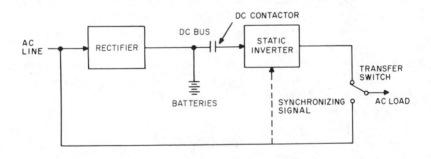

Fig 37
Oscillogram of Output Voltage of System in Fig 36 During Transfer

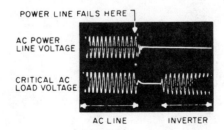

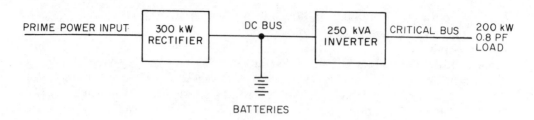

Fig 38
Nonredundant Uninterruptible Power Supply

During normal operation, the prime power and rectifier supply power to the inverter and also charge the battery that is *floated* on the dc bus and kept fully charged. The inverter converts battery power from dc-ac for use by the critical loads. The inverter alone governs the characteristics of the ac output, and any voltage or frequency fluctuations or transients present on the utility power system are completely isolated from the critical load.

In the event of a momentary or prolonged loss of power, the battery (which is floated on the dc bus) will supply sufficient power to the inverter to maintain its output for a specified time from a few minutes to several hours until the battery has discharged to a predetermined minimum voltage.

Figure 39 shows an oscillogram of the continuous-load voltage supplied by the system of Fig 38, even when the prime source of power fails.

Upon restoration of the prime power, the rectifier section will again resume feeding power to the inverter and will simultaneously recharge the battery. The rectifier is therefore sized 1.2-1.5 times larger than the inverter kilowatt output rating to account for this battery charging demand and inverter losses.

Static systems, as described, provide
(1) Precise uninterruptible power
(2) Low maintenance, no moving parts
(3) High efficiency, static conversion devices

System availability should be as high as economically justifiable and may be calculated by using the following formula with all figures in the same units, usually hours:

$$A = \frac{MTBF}{MTBF + MTTR}$$

where
A = system availability
$MTBF$ = mean time between failures
$MTTR$ = mean time to repair

A typical specification for an uninterruptible power supply system is given in Table 15.

Table 15
Typical Nonredundant 3φ UPS Performance Specifications

Input (Rectifier/Charger)	
Voltage	208 V or 480 V, ±10%, 3-phase
Power factor	Minimum 0.8 at rated load
Frequency	50 or 60 Hz, ±5%
Harmonic content of current	10% (5% preferred)
Startup current limiting	Maximum 25% of full-load current (energizing rectifier transformer with inverter at no load)
Startup "walk in"	15 to 30 s to full load
Steady-state current limiting	Adjustable, with two standard settings: 1) For utility power, 125% rated load 2) For emergency power, 100% rated load plus 5 kVA
Output (Inverter)	
Voltage	208 V or 480 V, 3-phase, 3- or 4-wire
Regulation	1) ±2% for balanced load 2) ±3% for 20% unbalanced load (100%, 80%, 80% or 100%, 100%, 80%)
Line drop compensation	0 to 5%, adjustable
Transient response	1) ±5% for loss or return of ac input power 2) ±8% for 50% load step 3) ±10% for bypass or return from bypass
Transient recovery	Return to steady-state conditions within 100 ms after a disturbance
Harmonic content of voltage	4% total, 3% any single harmonic
Phase displacement	1) $120° ± 1°$ for balanced load 2) $120° ± 3°$ for 20% unbalanced load
Frequency	50 or 60 Hz
Regulation	± 0.1 Hz
Line sync range	± 0.5 to 1.0 Hz, adjustable
Slew rate	Maximum 1 Hz/s
Current capability	
Overload	125% for 10 m and 150% for 10 s
Fault clearing	150% to 300% for 10 cycles, maximum limited for self-protection
DC Link (Battery)	
Battery type	Lead–acid or nickel–cadmium (NICAD)
Float voltage	Lead–acid 2.2–2.25 V/cell NICAD 1.4–1.42 V/cell
Equalize voltage	Lead–acid 2.35 V/cell NICAD 1.6 V/cell
End voltage	Lead–acid, minimum 1.6 V/cell NICAD minimum 1.1 V/cell (setting also determined by inverter input voltage window)
Recharge time	10 times discharge time
Energy storage capacity	Sized to requirement (normally 15 min)

Table 15 *(Continued)*

General Characteristics and Requirements	
3φ Output ratings	32.5 to 600 kVA at 0.8 power factor
Efficiency	77% to 90% (improves as kVA rating increases)
Dimensions and weight	Depends on kVA rating
Controls	Startup, emergency shutdown, synchronous transfer to bypass and all adjustment functions required for operation and maintenance
Meters	AC voltmeters and ammeters with phase selector switches for both input and output, DC voltmeter and charge/discharge ammeter
Alarms	Indicating 10 to 20 special conditions or malfunctions such as output over- and undervoltage, battery discharge, fan failure, auto bypass, etc
Environmental	
Ambient temperature	Within 0° to 40°C operating and −20° to 70°C nonoperating
Relative humidity	0 to 95% at any operating temperature
Reliability	MTBF 200 000 h minimum (includes available utility power via bypass)
Maintainability	MTTR 40 min maximum (when parts are on-site)
Available Options	
Frequency conversion	50 to 60 Hz or 60 to 50 Hz (only for redundant type UPS without bypass)
Expandability	Can be paralleled with like UPS modules
Electromagnetic interference suppression	Suppression of radiation on all sides and conducted on input, output, and control cables
Acoustical noise suppression	Maximum 76 dB at 5 ft from surface
Extended operating temperature capability	From 40°C to 50°C
Automatic battery equalizing charge	Activated and timed after each battery discharge
Circuit breaker motor operators	For input, output, and battery circuit breakers
Mimic bus	An illuminated one-line diagram indicating operational status
Remote status monitoring and alarm panel	Monitors special conditions and malfunctions up to 500 ft away
Additional meters	Input and output wattmeters, elapsed time and frequency meters, rectifier output dc ammeter
Special conditions to be identified by user	Damaging fumes Excessive moisture Excessive dust Abrasive dust Steam Oil vapor Explosive mixtures of dust or gases Salt air Abnormal vibration, shocks, or tilting Weather or dripping water Special transportation or storage conditions (user to identify method of handling equipment) Extreme or sudden changes in temperature Unusual space and weight limitations Unusual operating duty Unusually high system impedance

Table 15 *(Continued)*

Special conditions to be identified by user *(continued)*	Seismic considerations Electromagnetic fields Radioactive levels above natural background Abnormally high system voltages to ground Nonlinear load or one generating excessive harmonic or ripple current Inability for the dc source to accept a current in the reverse direction Acoustical noise limitations Type of battery or power supply provided by user

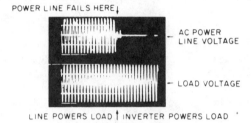

Fig 39
Oscillogram of System in Fig 38 With Powerline Failure

The nonredundant system shown in Fig 38 has the advantage of simplicity and low cost. Large systems can have a predicted reliability of 20 000 hours of mean time between failures using handbook reliability data; but field experience indicates that reliabilities on the order of 40 000 hours of mean time between failures can be achieved. The nonredundant system has the disadvantage of disturbing the critical bus in the event of an inverter failure. This disadvantage can be overcome by the use of a static bypass switch, as shown in Fig 42, or a redundant system, as shown in Fig 40.

5.4.4 Redundant Uninterruptible Power Supply. Figure 41 is an oscillogram of the two-inverter transfer system voltage during malfunction of one inverter similar to the arrangement shown in Fig 40. In the redundant system, each half of the system has a rating equal to the critical load requirements. The basic power elements (rectifier, inverter, and interrupter) are duplicated, but it is usually not necessary to duplicate the battery since its inherent reliability is extremely high. Certain control elements such as the frequency oscillator may also be duplicated.

The static interrupters isolate the faulty inverter from the critical bus and prevent the initial failure from starting a *chain reaction*, which might cause the remaining inverter to fail.

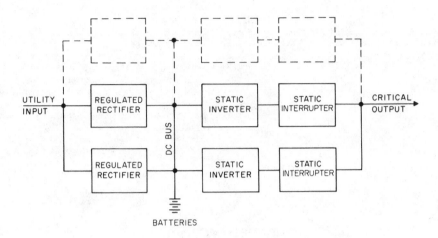

Fig 40
Redundant Uninterruptible Power Supply

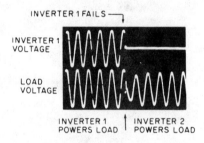

Fig 41
Oscillogram of Output Voltage of System in Fig 40 Upon Inverter Failure

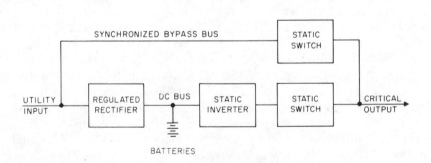

Fig 42
Uninterruptible Power Supply with Static Transfer Switch

Large systems that require multiple rectifier/inverters to handle the load requirements would require one additional rectifier/inverter to provide redundancy. Redundant systems should consist of the fewest parallel paths required to supply the critical load requirements, plus one additional path for redundancy. A larger number of smaller rated paths does not necessarily provide increased reliability since they add unnecessary additional components that are themselves subject to failure.

The cost of a redundant system is approximately $(N + 1)/N$ greater than that for a nonredundant system, where N equals the least number of parallel paths required for a nonredundant system. However, such a system is 2–4 times more reliable than a nonredundant system.

5.4.5 Nonredundant Uninterruptible Power Supply With Static Transfer Switch. An alternate method of increasing the overall system reliability is a static transfer around the faulted inverter as shown in Fig 42. When an inverter failure is sensed, the critical load can be transferred to the bypass circuit in less than 5 ms. Figure 43 is an oscillogram of the load voltage as supplied during a transfer of the power source.

The static transfer switch adds about 10% to the cost of a nonredundant system, but is 8–10 times more reliable.

5.4.6 Parallel Redundant Uninterruptible Power Supply. Figure 44 shows a parallel-supplied parallel redundant UPS system. Reliability is a paramount consideration in this 1000 kVA system. Solid-state sensors and static switches, not shown, are installed to clear a malfunctioning inverter without effect on critical computer load.

This parallel *load-sharing* configuration is usually specified when a static bypass cannot be effectively applied, such as when the utility input power quality is so poor that it cannot be relied upon for even short periods or when the utility input power is at a frequency different than that required by the load.

5.4.7 Cold Standby Redundant Power Supply. This configuration consists of two basic UPS systems, two static switches, and a common battery; see Fig 45.

Fig 43
Oscillogram of Static Switch in Fig 42 Load Voltage

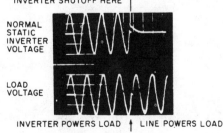

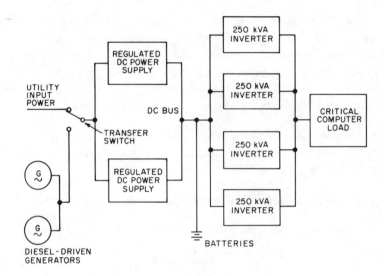

Fig 44
Parallel-Supplied Parallel Redundant Uninterruptible Power Supply

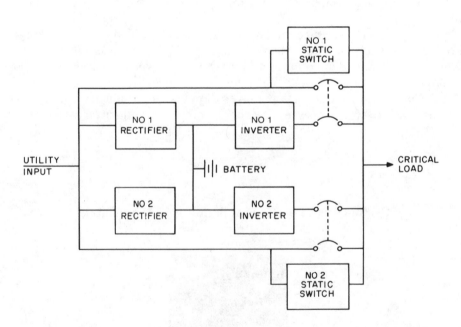

Fig 45
Cold Standby Redundant Uninterruptible Power Supply

One UPS operates on-line and the other is turned off. Should the operating UPS fail, its static bypass circuit will automatically transfer the critical load to the utility source without interruption of the load. The second UPS is manually energized and placed in the bypass mode of operation. To transfer the critical load to the second UPS external make-before-break, nonautomatic circuit breakers are operated, placing the critical load on the second UPS bypass. Finally, the critical load is returned from the bypass to the second UPS inverter via the static switch. The two inverters cannot be operated in parallel and, therefore, an interlock circuit must be provided to prevent this condition.

The cold standby configuration exposes the critical load to the utility line for short periods during transfer. Nevertheless, this configuration is found to be more reliable than the parallel redundant because the inverters are operated independently and single-point failures of load-sharing control circuits are eliminated. It is also more efficient to operate since one rectifier and inverter at 100% load is several percent higher in efficiency than two at 50% load.

Cold standby redundant costs about 5% more than the two-module parallel redundant configuration owing to the need for two static switches. However, this higher initial investment can be justified based on lower operating costs.

The cold standby redundant is typically specified instead of a nonredundant UPS system when the installation site is isolated and logistic support is poor. Of course the steady-state utility power must be acceptable to the critical load for short periods during transfer between modules.

5.4.8 Parallel Nonredundant Uninterruptible Power Supply With Static Bypass Switch. Figure 46 shows a parallel-supplied nonredundant uninterruptible power supply system. The installation consists of two discrete systems serving two computers. A synchronized bypass and static switch protect each load in the event of inverter fault. Should voltage be lost to a computer load, the static transfer switch will operate to reestablish voltage in less than one quarter of a cycle, fast enough to be considered continuous power for most loads.

5.5 Motor-Generators and Rotating UPS Systems

5.5.1 Introduction. Motor-generators are power systems that use a rotating ac generator to generate usable output power. If batteries are added to enable the system to continue operation without utility power input, then the system becomes an uninterruptible power supply (UPS).

Motor-generators use ac or dc motors to drive ac generators. They use their rotating inertia to *ride-through* input voltages, sags, or total loss for periods up to 500 thousandths of a second (500 ms). When a flywheel is added, the ride-through time can be many seconds.

For protection against loss of power lasting longer than 0.5 seconds, batteries or other energy storage devices should be added. Conversion of battery power to rotating energy is accomplished by one of two ways: (1) by use of a dc motor or (2) by use of a dc-ac solid-state inverter feeding an ac motor. To eliminate batteries, one system uses a diesel generator and a flywheel motor-generator. These systems are described here.

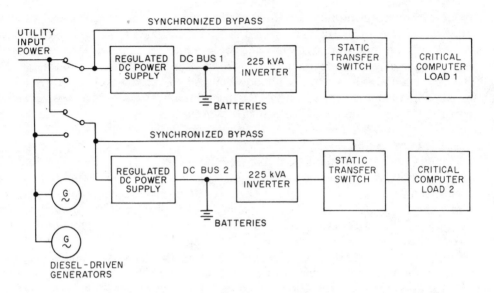

Fig 46
Parallel-Supplied Nonredundant Uninterruptible Power Supply

The ac generator is a versatile element, with a variety of means to drive it without interruption. AC generators have low harmonic content and distortion. They are capable of operating reliably at high temperatures and under severe short-term overloads. They have a field-proven history of high reliability and predictable failure. The bearings can be tested at periodic intervals to detect deterioration and allow for planned replacement.

All systems discussed in this chapter use brushless synchronous ac generators.

Three different motors may be used to drive the ac generator. They are ac induction (asynchronous), ac synchronous, and dc motors.

Induction motors offer the lowest cost per horsepower. Three-phase induction motors are most common. The output r/min is less than that from the ac synchronous motor because of *slip*. Torque increases as slip increases until the rotor stalls. Low-slip (0.4–0.7%) motors are recommended and available in sizes up to 100 hp. These motors have been designed to optimize slip, power factor, efficiency, and starting inrush. Typical slip is 0.4–0.7%, resulting in output of 1790 r/min under full load from a motor rated at 1800 r/min. Low-slip motors are sometimes built using oversized motors. These motors may have reduced efficiency.

Synchronous motors maintain a constant shaft speed independent of load and input voltage. They remain locked to the input frequency up to their pull-out torque. Their efficiencies are typically several percent higher than induction motors.

Synchronous motors have limited torque at speeds that are substantially lower than synchronus r/min. For this reason these motors are usually started under no-load conditions and are at operating speed when the load is applied. Synchronous motors offer power factor correction as an advantage in some applications.

DC motors offer the ability to regulate output frequency precisely. The speed of a dc motor may be varied by changing field excitation. For computer applications with tight frequency tolerance, a frequency regulator is used to adjust field excitation as input voltage and load vary. DC motors have brushes that wear over time and need replacement. Alarm systems are available that turn on a signal indicating the need for replacement. If brushes are regularly maintained, dc systems have the same life expectancy as induction or synchronous motors.

5.5.2 AC Motor-Generators. Nearly all ac motor-generators built today use synchronous generators. The motor may be either induction or synchronous. Induction motors have an inherent slip that causes the output frequency of the system to be less than the input by a fraction of a cycle. Synchronous motors duplicate the incoming frequency precisely and track it. Most computer-grade motor-generator sets utilize synchronous motors.

Several methods of construction are available, with the differences affecting ride-through time, isolation, and floor size (footprint). The smallest footprint is achieved by stacking the motor and generator and coupling their horizontal shafts using toothed pulleys and a belt (Fig 47), or by mounting the motor and generator on a common vertical shaft. The stacked horizontal units get added ride-through from the mass of the pulleys and belt.

Fig 47
Horizontal Coupling of Stacked Motor
and Generator Using Pulleys and Belt

Horizontal coupling can also be accomplished by building the motor and generator into a single frame or by coupling two separate frames on a base. Maximum isolation is attained by using separate frames, electrically isolated, and connected by an insulating coupling (Fig 48). Motor-generator sets with a common motor and generator in one rotor offer the smallest horizontal size but the least isolation of transients from input to output (Fig 49).

5.5.3 AC Motor-Generator with Flywheel. The addition of a flywheel adds mass and therefore increases the ride-through time of a motor-generator set. This

Fig 48
Horizontal Coupling of Motor and Generator Using
Two Separate Frames on a Base

Fig 49
Motor and Generator with Common Rotor

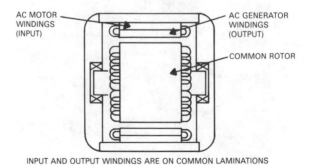

increased ride-through time is achieved at the expense of weight, size, efficiency, bearing life, and higher initial system cost. Air friction, or *windage*, increases rapidly as the radius of the flywheel and its speed increase. Bearing friction causes further losses, decreasing the overall system efficiency. The flywheel takes more input power to start up, so inrush current-limiting circuits are usually necessary. The addition of a flywheel also requires a larger motor for the extra horsepower necessary to restore r/min to the flywheel that is lost during input power sag or loss. Provision for the higher current input must be made.

Flywheel motor-generator sets are available with up to several seconds of ride-through time. Larger flywheels have proven to be impractical. It takes a mass of steel approximately 6 inches thick and 4 feet in diameter to maintain a 30 kVA load with less than 1 cycle frequency change for 0.5 seconds.

Maintenance of motor-generator sets consists of periodic checks for bearing noise. Motor-generator sets have a field-proven track record of very long MTBF (mean time between failures). The bearings are the most likely failure point. Their failure is predictable, so replacement prior to failure (typically once every 7–10 years) is the largest maintenance cost.

The construction of a motor-generator set can make bearing replacement easy or difficult. For example, some vertical motor-generators have the load support bearing on top to allow easy access. This allows the main bearing to be changed without pulling the rotor out of the case.

5.5.4 Battery/DC Motor/AC Motor-Generator Set. This UPS uses a standard motor-generator set for continuous power conditioning (see Fig 50). When utility power is normal, the dc motor is controlled by a field voltage control circuit to operate as a generator, maintaining the battery voltage. When utility power sags or fails, the dc motor is changed by a field voltage control circuit from a generator to a motor. It draws battery power to continue rotating the ac motor-generator without interruption.

Fig 50
Battery/DC Motor/AC Motor-Generator Set

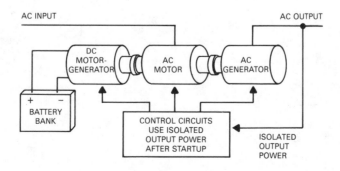

Advantages include

(1) Excellent power conditioning and load isolation
(2) Excellent overload and short-circuit capacity
(3) Maintenance and repairs do not require specially qualified personnel
(4) Failure of dc circuits or motor does not affect power conditioning

The disadvantage is

(1) DC motor brushes require maintenance and replacement

5.5.5 Battery/DC Motor/AC Generator. This UPS (Fig 51) uses a dc motor to turn the ac generator all the time. A rectifier charger converts the ac input to dc for the motor and battery. This system is less expensive than the one in 5.5.4, but it cannot continue to condition and isolate the powerline if any failure occurs. Some manufacturers of this type of system have separate circuits for the dc motor and the battery charger, and add a flywheel to ride through switchover time at utility power failure. Other manufacturers have simplified the system by reducing circuits and eliminating the flywheel.

Advantages include

(1) Output frequency can be regulated very closely at all times
(2) It conditions power and isolates the load from the utility
(3) Maintenance and repairs do not require specially trained personnel
(4) Cost is lower than systems with both an ac and dc motor
(5) The output frequency remains stable over varying input frequencies
(6) It may be used as a frequency changer

Disadvantages are

(1) Failure of the dc circuits prevents power conditioning
(2) Efficiency is reduced by rectifier circuits
(3) DC motor brushes require maintenance and replacement

Fig 51
Battery/DC Motor/AC Generator

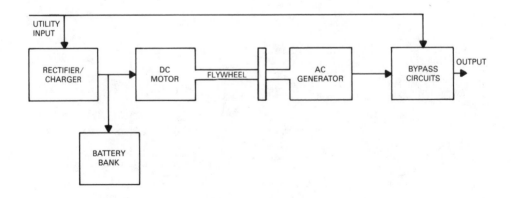

5.5.6 Off-Line Inverter/Motor-Generator System. This UPS system (Fig 52) substitutes a solid-state ac inverter for the dc motor in previous examples. The ac motor-generator set provides continuous powerline conditioning even if the dc and inverter circuits fail. Normally the inverter is off. At utility power loss, the inverter is turned on and converts stored battery energy to ac to drive the motor-generator set.

Advantages include

(1) Powerline conditioning and isolation are provided even if dc circuits fail

(2) Efficiency is high because the inverter is normally off

Disadvantages are

(1) An inverter may not be as reliable as a dc motor

(2) An inverter requires trained personnel for repair and maintenance

(3) Inverter failure may not be detected until it is turned on

5.5.7 On-Line Inverter/Motor-Generator System. This UPS system, called a *hybrid* UPS, shown in Fig 53, differs from *off-line* only in that the rectifier and inverter are always operating. The inverter is bypassed in case of failure.

Advantages include

(1) A failure in rectifier or inverter does not affect power conditioning

The disadvantage is

(1) An inverter is not as reliable as a dc motor

5.5.8 Engine/Motor-Generator System. This system (shown in Fig 54) consists of an internal combustion engine, an electromechanical clutch or a magnetic clutch, a horizontal axis flywheel, an ac generator, and an ac motor (plus control

Fig 52
Off-Line Inverter/Motor-Generator System

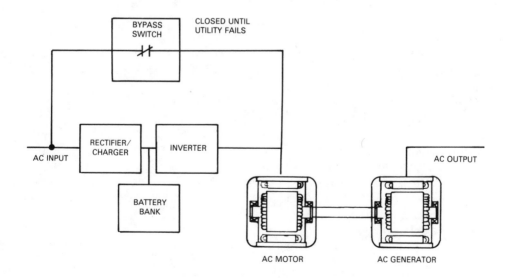

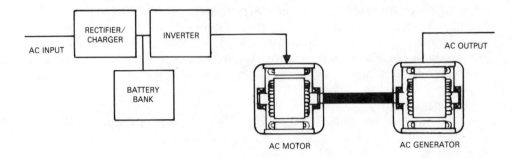

Fig 53
On-Line Inverter/Motor-Generator System

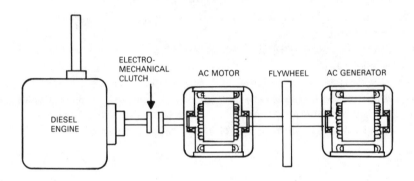

Fig 54
Engine/Motor-Generator System

and transfer equipment). In normal operation the utility input power to the ac motor drives the flywheel and ac generator, which in turn supplies power to the load. The flywheel stores kinetic energy.

When an interruption occurs in utility power, the flywheel drives the ac generator and starts the engine through the clutch. The engine then drives the ac generator, which continues to supply power to the load without interruption. With proper selection of components to minimize the start-up and run-up times of the diesel engine, the frequency dip can be kept to approximately 1.5-2 Hz without paying a premium. Thus, with a steady-state frequency of 60.0 Hz, the transient frequency would be from 58-58.5 Hz. The time for the diesel to start, come up to speed, and take the load is normally less than 2-3 seconds.

Advantages include

(1) Operating time after a utility power loss is limited only by fuel supply

(2) It offers excellent overload and short-circuit capacity

(3) Maintenance and repairs do not require specially qualified personnel

(4) The system has predictable failure modes

(5) There is no battery bank to maintain

Disadvantages are

(1) The mechanical clutch suffers wear and tear during each start and requires costly and frequent maintenance

(2) The engine crankshaft experiences high axial loads that require a special design; maintenance and replacement parts are costly

(3) An additional engine-generator may be needed to supply the air conditioning, lighting, and other similar loads if long-term operation is desired

5.5.9 Engine-Generator/Motor-Generator System. This system consists of an ac motor-generator set with a flywheel and an engine-generator set (Fig 55). In normal operation the utility input powers the ac motor-generator set, which in turn supplies power to the load. The flywheel stores kinetic energy.

When an interruption occurs in utility power, the flywheel energy supports the speed of the ac motor-generator set. A transfer switch disconnects the utility power from the motor-generator and connects the engine-generator output to the motor-generator. Simultaneously, the engine is started. As its speed increases, the output voltage rises, exciting the motor-generator set motor to act momentarily as a generator, assisting the engine in achieving full r/min.

Engine-generator sets under 100 kW in rating have been found to accelerate to full speed within 2 seconds. The backup generator has sufficient extra power to feed loads such as lights and air conditioning.

Advantages include

(1) Cost is far less than for both battery backup UPS and engine-generators

(2) Operating time during a utility power interruption is limited only by fuel supply

(3) It offers excellent overload and short-circuit capacity

**Fig 55
Engine-Generator/Motor-Generator System**

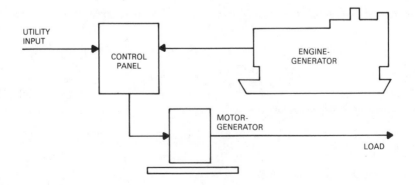

(4) The transfer of driving power to start the internal combustion engine is effected smoothly with electromagnetic action, avoiding clutching and destructive forces in the engine crankshaft

(5) Installation, service, repairs, and maintenance do not require specially qualified personnel

The disadvantage is

(1) Frequency is always below 60 Hz since an induction motor is used

5.5.10 Parallel Systems. Most systems described here can be connected in parallel for additional output power. In most cases, some added circuits are necessary to sense and adjust the phase angle of each ac generator output voltage to ensure they are synchronized when switching. An exception is the case of synchronous ac motors and the generators operating on the utility line. These motor-generator sets are always locked in synchronism. However, if two synchronous motor-generator sets are backed up by a dc battery and motor, control circuits are required to maintain synchronism during dc operation. Once connected together, synchronous ac generators will remain in synchronism.

5.5.11 Redundant Systems. Most systems described here may be connected to create redundant outputs. This may be accomplished by parallel operation at less than half power per system, or by connection through a sensing and switching circuit.

Induction motor-generator sets are more difficult to parallel than synchronous sets due to the continuous slip of the induction motor. With one unit loaded, the other unit being brought on-line must be loaded with an external load bank until the two generator outputs are in phase. Only then may the outputs be connected to each other to supply the load in parallel.

5.5.12 Bypass Circuits. UPS systems that have the same frequency and voltage input and output are usually supplied with a bypass circuit. This circuit consists of a static transfer switch or an electromechanical transfer switch, or both. The purpose of the bypass circuit is threefold. First, to connect the utility to the load in case of failure of the UPS. Second, to bypass the UPS for maintenance. Third, to assist the UPS output in driving high surge currents when faults occur. Most motor-generators and motor-generator-based UPS systems do not require ac bypass to the utility to clear faults. They have enough overload capacity to clear the faults without assistance.

5.6 References. The following publications shall be used in conjunction with this chapter.

[1] ANSI C84.1-1982, American National Standard Voltage Ratings for Electric Power Systems and Equipment (60 Hz).

[2] ANSI/IEEE C37.95-1973 (R 1980), IEEE Guide for Protective Relaying of Utility-Consumer Interconnections.

[3] ANSI-IEEE Std 100-1984, Dictionary of Electrical and Electronics Terms.

[4] ANSI/IEEE Std 241-1983, IEEE Recommended Practice for Electric Power Systems in Commercial Buildings.

[5] ANSI/IEEE Std 450-1987, IEEE Recommended Practice for Maintenance, Testing, and Replacement of Large Lead Storage Batteries for Generating Stations and Substations.

[6] ANSI/IEEE Std 485-1983, IEEE Recommended Practice for Sizing Large Lead Storage Batteries for Generating Stations and Substations.

[7] ANSI/NFPA 70-1987, National Electrical Code.

[8] ANSI/NFPA 101-1985, Life Safety Code.

[9] EGSA 101E-1984, Glossary of Standard Terms — Electrical.

[10] EGSA 101M-1984, Glossary of Standard Terms — Mechanical.

[11] NECA Electrical Design Library Series, vol 17, Electrical Design Guidelines, 1971.[15]

[12] NECA Electrical Design Library Series, vol 3/74, Emergency and Standby Power Generation, 1974.

[13] HEISING, C. R. and JOHNSTON, J. F., JR. Reliability Considerations in Systems Applications of Uninterruptible Power Supplies. *IEEE Transactions on Industry Applications*, vol IA-8, Mar/Apr 1972, pp 104–107.

[14] IEEE Committee Report. Reliability of Electrical Equipment, Pt 1. *IEEE Transactions on Industry Applications*, vol IA-10, Mar/Apr 1974, pp 213–235.

5.7 Bibliography

[B1] The Automatic Transfer Switch: Heart of Emergency Power — A Reliability Study of a Power Supply System — The Battery World. *Electrical Consultant*, vol 88, Nov 1972.

[B2] The Exciting World of Rechargeable Batteries. *Factory*, Apr 1967, pp 84–87.

[B3] GRIFFITH, D. C. and YUEN, M. H. Static No-Break Power for Critical Loads in a Modern Oil Refinery. *Conference Record of the 1967 IEEE Industry and General Applications Group Annual Meeting*, IEEE 34C62, pp 643–652.

[B4] GROSS, S. Rapid Charging of Lead Acid Batteries. *Conference Record of the 1973 IEEE Industry Applications Society Annual Meeting*, IEEE 73CH0763-3IA, pp 905–912.

[B5] HAUCK, T. A. Motor Reclosing and Bus Transfer. *IEEE Transactions on Industry and General Applications*, vol IGA-6, May/June 1970, pp 266–271.

[B6] HELMICK, C. G. Designing for System Reliability in Large Uninterruptible Power Supplies. *Conference Record of the 1971 IEEE Industry and General Applications Group Annual Meeting*, IEEE 71C1-IGA, pp 371–384.

---

[15] NECA publications can be obtained from the National Electrical Contractors Association, Accounting Department, 7315 Wisconsin Ave, Bethesda, MD 20814.

[B7] HELMICK, C. G. Uninterruptible Power Supply Systems — What, Why, Where, and When? Presented at the 34th American Power Conference, Chicago, IL, Apr 18-20, 1972.

[B8] KATZAROFF, P. A Base Guide to Uninterruptible Power Systems. *Conference Record of the 1974 IEEE 26th Annual Conference of Electrical Engineering Problems in the Rubber and Plastics Industries*, IEEE 74CH0831-8IA, pp 1-6.

[B9] KUSKO, A. and GILMORE, F. E. Application of Static Uninterruptible Power Systems to Computer Loads. *Conference Record of the 1969 IEEE Industry and General Applications Group Annual Meeting*, IEEE 69C5-IGA, pp 635-639.

[B10] KUSKO, A. and GILMORE, F. E. Concept of a Modular Static Uninterruptible Power System. *Conference Record of the 1967 IEEE Industry and General Applications Group Annual Meeting*, IEEE 34C62, pp 147-153.

[B11] LAWSON, L. J. New Uninterruptible Power System Alternatives Using High Capacity Kinetic Energy Wheels. *Conference Record of the 1973 IEEE Industry Applications Society Annual Meeting*, IEEE 73CH0763-3IA, pp 151-156.

[B12] LAWSON, L. J. A True No-Break, Off-Line Uninterruptible Power Supply. *Conference Record of the 1967 IEEE Industry and General Applications Group Annual Meeting*, IEEE 34C62, pp 154-158.

[B13] RELATION, E. A. UPS Systems for Critical Power Supplies. *Conference Record of the 1971 IEEE Industry and General Applications Group Annual Meeting*, IEEE 71C1-IGA, pp 877-884.

[B14] RELATION, E. A., WINPISINGER, J. L., and MITCHELL, J. T. Uninterruptible Power System Using an Improved Magnetic Voltage Stabilizer. *Conference Record of the 1973 IEEE Industry Applications Society Annual Meeting*, IEEE 73CH0763-3IA, pp 17-23.

[B15] RENFREW, R. M. Successful Uninterruptible Power Systems for Computers. *Conference Record of the 1968 IEEE Industry and General Applications Society Annual Meeting*, IEEE 68C27-IGA, pp 787-792.

[B16] ROBERTS, A. M. Power Failure Ride-Through for an Inverter System Using Its Own Induction Motor Load as the Energy Source. *Conference Record of the 1968 IEEE Industry and General Application Group Annual Meeting*, IEEE 68C27-IGA, pp 737-742.

[B17] SCHWARM, E. G. and LITTLE, A. D. Computer Uninterruptible Power System with High Speed Static Bypass. Presented at the Summer Power Meeting and International Symposium of High Power Testing of the IEEE Power Engineering Society, Portland, OR, July 18-23, 1971.

[B18] SUMMERS, G. E. Providing Reliable Power for Computer Systems. *Plant Engineering*, Jan 7, 1971.

[B19] System for Orderly Emergency Shutdown. *Modern Manufacturing*, Dec 1969.

[B20] Uninterruptible Power System Prevents Computer Downtime. *Rubber World*, Nov 1970, pp 58–60.

[B21] TERVAY, J. C. Nickel–Cadmium Pocket Plate Batteries for Standby Power Applications and Systems. Nife, Inc, 23 Dixon Avenue, Copiague, NY 11726.

[B22] WALKER, L. H. Inverter for UPS with Subcycle Fault Clearing Capabilities. *Conference Record of the 1971 IEEE Industry and General Applications Group Annual Meeting*, IEEE 71C1-IGA, pp 361–370.

[B23] WOLPERT, T. Uninterruptible Power Supply for Critical AC Loads — A New Approach. *Conference Record of the 1973 IEEE Industry Applications Society Annual Meeting*, IEEE 73CH0763-3IA, pp 595–602.

Chapter 6
Protection

6.1 Introduction. This section is intended to discuss recommended practices and guidelines in applying protection to emergency and standby power systems. Even though standard practice for protection of equipment should always be given full consideration when applying the equipment in emergency and standby use, reliability often warrants special consideration when designing a power supply for critical loads. All national, state, and local codes and standards applicable to protection of the components that make up the emergency or standby power system are a necessary part of the design process. It is not the intent of this chapter to list all applicable codes and standards, but some pertinent ones are referenced where necessary to aid this guideline.

Protection for individual components that make up the most common emergency and standby power systems is discussed, with emphasis on maintaining the required integrity and reliability of the system. Proper application and maintenance of systems are discussed in other chapters and should be considered an important aspect of protection.

6.2 Short-Circuit Current Considerations. Of the many areas of concern for protection of components that make up an emergency or standby power system, one needing special consideration is that of magnitude and duration of short-circuit current from the emergency or standby power source. Fault conditions obviously have a direct effect on the availability of the power supply to serve its intended purpose. Studies should be made to determine available short-circuit current throughout the system supplied by an emergency or standby power supply, especially at switching and current-interrupting devices.

In evaluating the performance of an emergency or standby generator under fault conditions, a critical concern is whether or not sufficient fault current is available for sufficient duration to selectively trip overcurrent devices in a properly coordinated system. In most cases, emergency or standby power sources, unless derived from an electric utility, do not have as much fault current available as do normal base-loaded power systems. When both sources are designed to supply a distribution system, through automatic or manual switching devices, the magnitude of the

171

fault current available from the normal supply determines the required interrupting or withstand rating, or both, of the system components. The difficulty lies in designing optimum selectivity and coordination for both power sources. An emergency or standby generator should be evaluated as to whether or not it will supply enough fault current to trip a branch overcurrent device that is coordinated with a larger normal power source overcurrent device.

Selective coordination of overcurrent devices is the process of applying these devices so that one will operate before another under given levels of fault currents, thereby allowing effective isolation of the faulted circuit from unfaulted circuits. For a detailed analysis of the principals of selective coordination, see ANSI/IEEE Std 242-1986 [8].[16]

Evaluating the available fault current from an emergency or standby generator includes determining the magnitude of the fault current and how long the fault current will exist and how it may change. Test information on generator short-circuit characteristics should be requested from the manufacturer. A critical mistake is to assume that the speed of the generator will remain unchanged for a fault fed only by the generator. The coordination of the mechanical and electrical stored energy in a generator set and the type of excitation system will determine how long the engine and generator will sustain fault current and how rapidly this fault current will decay. In some cases, the initial power to a fault is higher (sometimes several times higher) than the prime mover power rating. The result can be a rapid reduction in speed. This overloading of the prime mover will occur more readily in small generator set applications than in larger power systems since the X/R ratios of both the generator windings and distribution circuits are lower.

An excitation system cannot respond fast enough to affect the first one or two cycles of fault current so, at this point, the type of excitation system does not matter. So, if the fault is not cleared in the first 2–4 cycles, the next concern is whether or not the generator can recover, after the fault is cleared, to continue service to unfaulted branch circuits.

Figure 56 illustrates the rate of decay of three-phase fault current from tests on three typical generators, the first two of which are separately excited (with fixed excitation) and tested at no load. The curves are approximations of actual test results. In Fig 56(a) about 40 cycles elapse before the fault current decays to the full-load rating of the generator. In Fig 56(b), the fault current decayed to the full-load rating in about 25 cycles. The current (I) shown is in per unit (pu) based on generator full-load current rating.

Figure 56(c) illustrates an approximation of short-circuit current of a shunt-excited generator tested with a three-phase fault, and without short-circuit sustaining capability. In this case, the current has decayed to full-load rating in 10 cycles and to nearly zero in 20 cycles.

When two commonly used molded-case circuit breakers are applied in series in a system with available fault current as shown in Fig 56, even though the upstream

---

[16] The numbers in brackets correspond to those of the references listed at the end of this chapter; when preceded by B, they correspond to the bibliography at the end of this chapter.

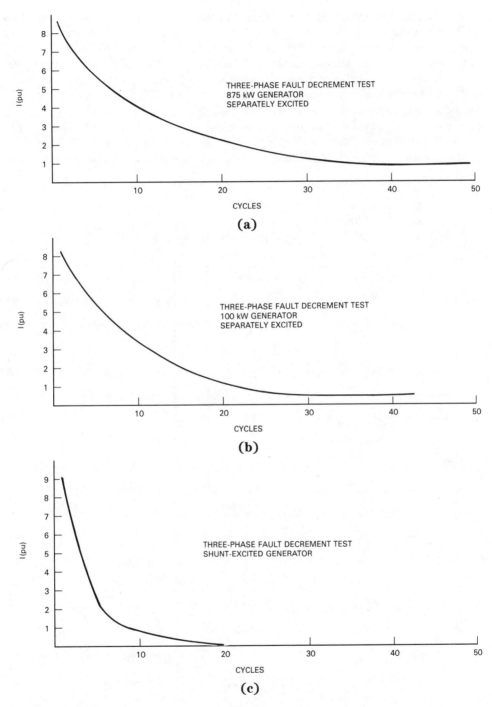

**Fig 56
Three-Phase Decrement Curves for Engine-Generators**

breaker is substantially larger, selective instantaneous tripping is not assured on faults below the downstream breaker unless the fault current at this point is higher than the maximum clearing time of the instantaneous trip level of the branch breaker but less than the minimum instantaneous trip level of the larger upstream breaker. In all three parts of Fig 56 the fault current decays rapidly but, depending on the instantaneous setting of the larger upstream breaker, the possibility exists for its unlatching level to be exceeded. If this happens, both breakers will trip and power is lost to unfaulted branch circuits. This concept is illustrated in Fig 57. The figure shows a case where the fault current, taken from Fig 56(b), is above the maximum clearing level of the branch breaker but is also above the minimum

Fig 57
Three-Phase Short-Circuit Current versus
Molded-Case Circuit Breakers in Series

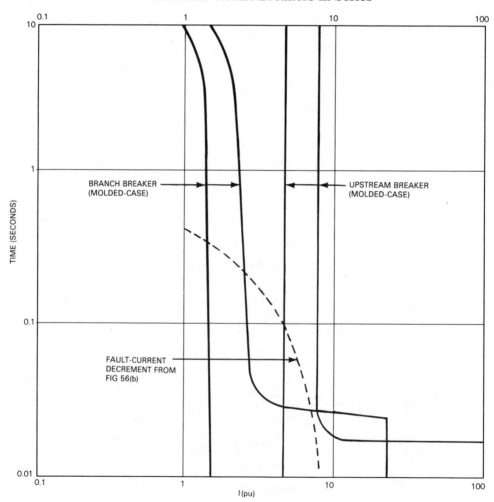

clearing level of the larger breaker. For this case, the unlatching characteristic of the larger breaker must be known to determine whether or not it will trip. It should be pointed out also that the illustrated fault current was measured at the generator terminals and presented in Fig 57 for convenience. The maximum fault current from the same generator will actually be less in the branch circuit (and possibly decay at a slower rate) than the illustration shows, depending on the impedance between generator terminals and fault. An important point, however, is that when evaluating whether or not a breaker will trip for cases where the margin between breaker tripping characteristics is close, the unlatching time of the larger breaker and not its clearing time may be the determining factor in whether or not selective coordination is achieved.

For the cases illustrated in Fig 56, only three-phase fault conditions are shown. In actual practice, unbalanced faults are much more common, especially line-to-ground in grounded systems. For this reason, evaluating only three-phase fault conditions in the application of an engine-generator is insufficient. Shunt exciter systems may derive their power from line-to-line, line-to-neutral, or from all three phases. Depending on the particular design, it is possible that faults on line-to-line or line-to-neutral may not decay to a self-protecting level as shown in Fig 56. The degree to which this can be a problem should be determined on a case-by-case basis. It is possible for a generator and excitation system to be designed to assure collapse of the generator terminal voltage and fault current on any kind of fault at the generator terminals. Generator and excitation systems are also available to assure a predicted level of fault current. A user should specify the type of system that best fits his needs.

Figure 58 is an approximation of an actual test on a turbine-driven generator with and without excitation support under three-phase fault conditions. The curves show how little influence the excitation system has in the first few cycles, but the sustained fault current is substantially different from that without excitation support.

Fig 58
Three-Phase Decrements for 900 kW Turbine Generator

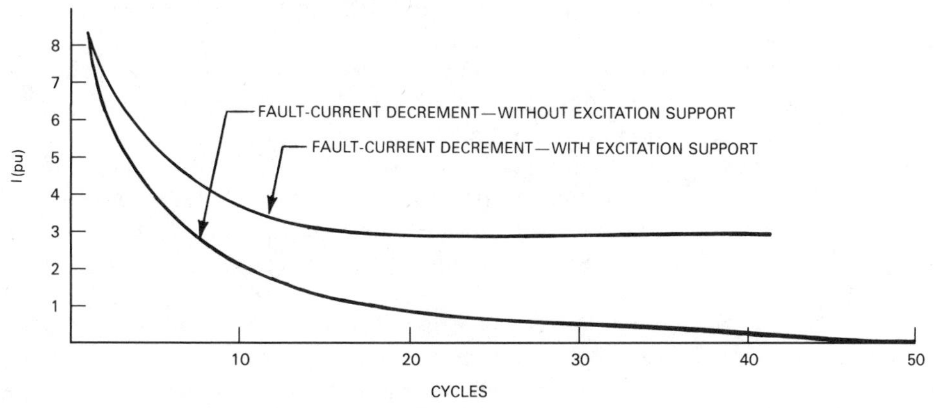

This type of fault-sustaining support will produce even higher than 3 pu fault current for line-to-line and line-to-ground faults. These values are typically 5 pu and 8 pu, respectively.

6.3 Transfer Devices

6.3.1 Codes and Standards. Several codes and standards that relate to transfer devices are referenced throughout this section. These are listed in 6.11 as: [1], [2], [3], [4], [6], [9], [12], [13], [14], and [15].

Other standards not referenced in this section that may serve as useful references are listed in 6.12 as: [B2], [B3], and [B4].

Codes and standards applicable to protection of transfer devices vary somewhat between different usages of the equipment. ANSI/NFPA 70-1987 [12] (National Electrical Code [NEC]), for instance, is limited in its requirements for transfer equipment in emergency and standby power applications. ANSI/UL 1008-1983 [14] contains information on construction and testing of automatic transfer switches. Most users today place considerable emphasis on adequacy of equipment as determined by recognized testing labs. For this reason, testing has been given special attention in this section. ANSI/UL 1008-1983 [14] originated in 1972, and only those automatic transfer switches that can comply with the required load duty and fault-current withstand ratings are eligible for listing. The primary concern of testing labs, however, is safety, and the user should recognize reliability considerations in the proper selection and application of a transfer switch. It should be noted that CSA C22.2/178-1978 [15], the Canadian standard on automatic transfer switches, is similar to ANSI/UL 1008-1983 [14]. Also, because of worldwide need, the International Electrotechnical Commission (IEC) is currently preparing a standard on automatic transfer switches.

6.3.2 Current Withstand Ratings. Transfer switches, being a vital part of the proper operation of the system and because of their application requirements, should be given special consideration over normal branch-circuit devices. The design, normal duty, and fault-current ratings of the switch play an important part in its application and protection scheme. It must be capable of closing into high inrush currents, of withstanding fault currents without damage, and of severe duty cycle in switching normal-rated load. All are capability considerations and thus important to protection, but emphasis in this chapter will be mainly on fault withstandability. The coordination of overcurrent protection devices with transfer switch ratings, under fault conditions, is one of the most important aspects of providing reliable operation of a standby or emergency power system.

The destructive effects of high fault currents consist mainly of two components: (1) magnetic stresses that attempt to pry open the switch contacts and bend bus bars, and (2) heat energy developed that can melt, deform, or otherwise damage the switch. Either or both of these components can cause switch failure.

A fault involving high short-circuit currents usually causes a substantial voltage drop that will be sensed by the voltage sensing relays in the automatic transfer switch. It is imperative for protection that the switch contacts remain closed until protective devices can clear the fault. Separation of contacts, prior to protective device operation, can develop enough arcing and heat to damage the switch. Normally employed time delay to prevent immediate transfer of mechanically held mechanisms and contact structures specifically designed with high contact pressure, in some cases utilizing electromagnetic forces to increase contact pressure, combine to provide the reliability and protection necessary in automatic transfer switch operation. It follows that proper application of the switch within its withstand rating is important to prevent contacts from welding together and to prevent any other circuit path joints and connections from overheating or deforming, thus prolonging the life and increasing reliability of the switch.

ANSI/UL 1008-1983 [14] presents specific test requirements to ensure the withstand capabilities of switches. Included are methods and types of overcurrent device application, minimum available short-circuit currents required, and allowable damage criteria. Also, power factor requirements of the test circuit are given. Manufacturers should be consulted to determine the method of testing applied to transfer switches. Determining whether a fuse or circuit breaker (and what size and type) was used, and determining the X/R ratio of the test circuit are both important aids in judging whether or not the switch is suitable for the actual application. Current-limiting fuses, for example, would considerably limit the duration of short-circuit current compared to the application of circuit breakers. This is discussed further in 6.3.4. Specifications should be carefully examined to recognize when asymmetrical and instantaneous peak fault currents are used to avoid being misled by specifications giving seemingly high numbers that may actually be asymmetrical or peak current ratings. Normally, symmetrical rms amperes should be used when coordinating time–current characteristics of switches and protective devices.

Additionally, ANSI/UL 1008-1983 [14] establishes withstand ratings for transfer switches using either integrally designed overcurrent protection or external overcurrent protective devices. In the case where circuit breakers are integrally incorporated into the design, it should be noted that the transfer switch contacts will not remain closed for the duration of a short circuit, but instead will interrupt the current. Thus the withstand rating can be considered the same as the interrupt rating. This interchange of the terms *interrupting* and *withstand* can lead to confusion, especially when transfer switches of the integral circuit breaker type with the trip units removed are used with external overcurrent protective devices. Caution is advised when evaluating the withstand rating for this case, since eliminating the trip unit can reduce the interrupting rating and therefore the withstand rating by 2 or 3 times.

Table 16 shows an example of a manufacturer's switch withstand current ratings when protected by current-limiting fuses or circuit breakers. This table illustrates the difference in withstand current ratings for switches, owing to the inherent differences between breakers and fuses. The normal and emergency contacts have the same withstand ratings even though the normal source is usually capable of delivering higher fault currents. The table is not intended as a guide for all switches as the data will vary among manufacturers and types of switches.

6.3.3 Significance of X/R Ratio. It is the X/R ratio of a circuit that determines the maximum available peak current and thus the maximum magnetic stresses

Table 16
Example Withstand Current Ratings for Automatic Transfer Switches

	When Used with Class J and L Current-Limiting Fuses		When Used with Molded-Case Breakers	
Switch Rating (amperes)	Withstand Current Rating	Maximum Fuse Size (amperes)	Withstand Current Rating	Maximum Breaker Size (amperes)
30	100 000	60	10 000	50
100	100 000	300	22 000	150
260	200 000	400	22 000	600
400	200 000	800	35 000	600
600	200 000	1200	42 000	2500
800	200 000	1200	42 000	2500
1000	200 000	2000	65 000	2500
1200	200 000	2000	65 000	2500
1600	200 000	3000	85 000	2500
2000	200 000	3000	85 000	2500

Available Symmetrical Amperes rms at 480 V AC and X/R Ratio of 6.6 or Less

NOTE: X/R ratio, size of overcurrent protective devices, and withstand current ratings vary depending upon the manufacturer.

that can occur. As the X/R ratio increases, both the fault withstandability of the switch and the capability of an overcurrent protective device become more critical. The power factor, X/R ratios, and their relationships to peak current can be found in Table 2 of ANSI/IEEE C37.26-1972 [3]. In many instances a circuit breaker symmetrical current interrupting rating or a transfer switch withstand rating is reduced if applied at an X/R ratio greater than what the device safely withstood at test.

When current-limiting fuses are employed as protective devices for switches, the peak instantaneous let-through current passed by the fuse should be equal to or less than the instantaneous peak current for which the switch has been tested. Also, the fuse interrupting rating and test X/R ratio should be greater than the circuit available fault current and rated X/R ratio, respectively. Tables 17–19 show how X/R ratios vary for circuit breaker interrupting ratings, fuse interrupting ratings, and transfer switch withstand ratings.

6.3.4 Withstand Ratings with Respect to Time. Fault-current magnitude and time of duration determine the heat energy and thermal stress developed during a fault, and this energy is normally designated as I^2t. In addition to current magnitude withstand ratings, transfer switches have I^2t ratings that, if exceeded, can damage or possibly even destroy a switch.

Available I^2t at a switch will vary with the magnitude of fault current and the clearing time of the overcurrent device protecting the switch. To know the available I^2t, the type of overcurrent protective device must be known since different protective devices allow different magnitudes of fault-current let-through and have

Table 17
Molded-Case and Power Circuit Breaker Interrupting Requirements

Molded-Case Breakers	Power Breakers	Test Power Factor (%)	X/R Ratio
10 000 or less	—	45–50	1.73–1.98
10 001–20 000	—	25–30	3.18–3.87
20 001 and more	—	15–20	4.9–6.6
—	All ratings	15 maximum	6.6 minimum

Interrupting Rating (symmetrical amperes)

Sources: ANSI/IEEE C37.13-1981 [2] and ANSI/UL 489-1985 [B10].

Table 18
Automatic Transfer Switch Withstand Requirements

Withstand Test Available Current (symmetrical amperes)	Test Power Factor (%)	X/R Ratio
10 000 or less	45–50	1.73–2.29
10 001–20 000	25–30	3.18–3.87
20 001 and more	20 maximum	4 minimum

Source: ANSI/UL 1008-1983 [14].

Table 19
Fuse Interrupting Test Requirements

Fuse Class	Interrupting Test Current (symmetrical amperes)	Test Power Factor (%)	X/R Ratio
H*	10 000	45–50	1.73–1.98
K	50 000	20 maximum	4.9 minimum
	100 000		
	200 000		
J	200 000	20 maximum	4.9 minimum
L	200 000	20 maximum	4.9 minimum
R	200 000	20 maximum	4.9 minimum
T	200 000	20 maximum	4.9 minimum

Sources: ANSI C97.1-1972 [B1] and ANSI/UL 198B-1982 [B5], 198C-1981 [B6], 198D-1982 [B7], 198E-1982 [B8], and 198H-1982 [B9].
*When rated above 100 A.

different clearing times. Clearing time of a fuse will differ from that of a breaker, and each will differ among different specific types. Current-limiting fuses introduce an additional parameter to be considered, called peak let-through current. When applying current-limiting fuses, the time factor reduces to a fraction of a cycle.

ANSI/UL 1008-1983 [14] permits a transfer switch to be tested without an overcurrent protective device so long as the time of the test is at least as long as the opening time of a specified protective device at a specified level of fault current. Although the switch under these conditions may be capable of withstanding several cycles of the specified fault current, it may not be labeled as rated for this number of cycles. In actual applications, it is not uncommon for switches to be required to carry fault current for several cycles. The user is still bound to assure that the available fault current does not cause the I^2t rating of the switch to be exceeded.

Table 20 shows typical transfer switch ratings in I^2t as related to different peak let-through currents of fuses. These values will vary among different switch manufacturers.

6.3.5 Transfer Switch Dielectric Strength. In addition to current-carrying capabilities, transfer switches must be able to withstand voltage surges to satisfy their reliability requirements. Control devices in an automatic transfer switch initiate the operation of transfer or retransfer and thus heavily affect the reliability of the switch under normal conditions. However, these control devices do not have the physical size and dielectric space inherent in the main load current-carrying parts of the switch. For this reason, the user should ensure that high-quality products suitable for emergency equipment use are employed in the switch.

ANSI/UL 1008-1983 [14] requires dielectric voltage withstand tests of 1000 V plus twice-rated voltage. But the requirements are unclear as to what specific devices are to be tested and whether or not control devices are to be included. ANSI/UL 1008-1983 [14] requirements are intended mainly for safety, but to satisfy the reliability requirements of an emergency or standby power system, the user should consider additional surge protection depending on the exposure of the switch to voltage surges. Some considerations helpful in protecting the dielectric strength of a switch are

(1) Arc breaking capability to minimize flashover between sources and deteriorating of dielectric.

(2) Contact construction to minimize heat generated at high currents.

(3) Readily accessible contacts and components for easy visual inspection and replacement.

Some common causes of voltage transients in ac systems that might affect transfer switches are switching inductive loads, energizing and de-energizing transformers, and lightning and commutation transients. In most cases, it is impossible to eliminate all the causes of transients, and thus the transients themselves, so the next step is to assume they will occur and then take measures to protect sensitive equipment. It is recommended that transfer switches meet impulse withstand voltage test requirements as designated in ANSI/NEMA ICS 1-1983 [9] and voltage surge withstand capability as designated in ANSI/IEEE C37.90-1978 [4]. This is particularly important if solid-state voltage and frequency sensing is used.

Table 20
Typical Transfer Switch Characteristics When Used with Fuses

Switch Rating (amperes)	Available rms Symmetrical Fault Current (amperes)	Peak Let-Through Current (amperes)	I^2t (amperes^2-seconds)
125	200 000	26 000	600 · 10
225	200 000	45 000	2500 · 10
400	200 000	80 000	12 000 · 10
600	200 000	100 000	12 000 · 10
1000	200 000	150 000	35 000 · 10

ANSI/IEEE Std 141-1986 [6] provides an in-depth analysis of causes and effects of various kinds of overvoltages.

6.3.6 Protection with Circuit Breakers. Principals of coordination and selectivity between the breakers and devices to be protected in the distribution system follow those of most typical systems (assuming adequate fault-current availability). Reliability of power to critical loads will depend on the selectivity between breakers. Two areas of concern regarding breakers and transfer switches are evident: protecting the switch according to its withstand rating and, at the same time, achieving proper selectivity for service reliability. Selectivity becomes more of a problem when transfer switching and overcurrent protection are combined as an integral unit. The integral protective device setting must be known to allow coordination with external overcurrent devices.

Figures 59 and 60 show a simple illustration of a typical small system protected with circuit breakers. Complete coordination is achieved by employing time delay in the short time region on breakers A and B. The transfer switch is protected for any fault downstream of the switch by breakers B and C. In this case, the capability of the cable associated with the transfer switch is the limiting factor and reduces the apparatus withstand rating. The switch by itself would have a higher rating. Also, in achieving coordination, the switch and associated wiring are required to withstand as much as 6 cycles of fault current for a fault between the switch and the branch breakers before the fault current is cleared. The rating of this switch will accommodate this requirement if the fault current does not exceed 20 000 A. An assumption in this illustration is that the generator will provide sufficient fault current to trip a branch breaker for a fault below one of these breakers. In most instances, available fault current from an emergency or standby generator will be substantially less than that of the normal source. When this is the case, the emergency or standby source should be located as close as possible to the critical load from a distribution standpoint to minimize the number of coordination levels and breaker sizes. As previously stated, proper selectivity of breaker operation is no different from that in most distribution systems. However, by investing in a breaker with a time-delay trip feature, an additional advantage is gained in the selectivity in preventing nuisance starting of and transfer to the emergency or standby source.

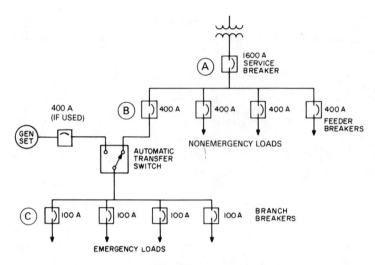

Fig 59
Emergency Power System with All Circuit Breaker Protection

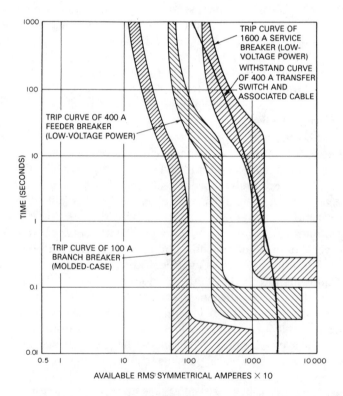

Fig 60
Coordination Chart of Emergency Power System
with All Circuit Breaker Protection

When applying circuit breakers for protection, the transfer switch I^2t rating must correlate with the maximum clearing time of the breaker protecting it even with instantaneous tripping. For instance, it is possible for a transfer switch to have a withstand rating of 100 000 symmetrical amperes when used with a specific current-limiting fuse, but may only have a withstand rating of 30 000 symmetrical amperes when used with a specific circuit breaker with an instantaneous trip. The same reasoning applies for breakers with different fault-clearing times, especially considering time-delayed trips versus instantaneous trips. ANSI C37.16-1980 [1] recognizes this by requiring reduced short-circuit interrupting ratings of circuit breakers for short-time delay applications. For example, a 225 A frame low-voltage power circuit breaker, with an instantaneous trip unit, has a 22 000 symmetrical amperes interrupting rating at 480 V. The same breaker without an instantaneous trip has a reduced rating of 14 000 symmetrical amperes at 480 V.

It is often difficult to predict future expansion with the present emphasis on cost-effective systems. It is especially difficult to justify added capital expenditures for speculated load growth. A situation not so unusual is for available short-circuit current to eventually exceed breaker and transfer switch interrupting and withstand ratings when the normal supply transformer bank is increased in rating to accommodate unexpected load growth. A solution is to apply current-limiting fuses in line with the existing breakers and the equipment whose rating has been exceeded. The fuses should be applied in accordance with recommendations from the circuit breaker manufacturers. This offers an economical compromise of limiting short-circuit current and I^2t let-through and still maintaining some operating flexibility. However, coordination can be compromised. An alternative solution is to apply current-limiting reactors whereby overcurrent device coordination can be maintained.

6.3.7 Protection with Fuses. First cost often favors fuse application over circuit breakers. Another advantage is that fuses can safely interrupt higher short-circuit currents than breakers and with faster clearing times. A disadvantage is the requirement to replace fuses after fault clearing. Also, fuses are nonadjustable and sometimes a compromise in final coordination in the power system is necessary when an ideal fuse rating is unavailable. Gang operation of poles in a circuit breaker offers an advantage over fuses by eliminating possible single-phase conditions.

When fuses are used, peak let-through current and I^2t energy let-through should be coordinated with the same characteristics of the transfer switch to be protected. These characteristics vary among fuse manufacturers and types of fuses and, therefore, the manufacturer should be consulted for each particular fuse considered. Additionally, transfer switches may be rated for operation in series with a specific fuse for which they were tested. If another class of fuse with the same ampere rating and interrupting rating is substituted, the transfer switch could possibly fail under fault conditions if the substitute fuse permits a higher peak current and I^2t energy let-through. The same reasoning applies when comparing a current-limiting fuse to a circuit breaker equipped with an instantaneous trip with a clearing time as fast as 1½ cycles. The circuit breaker will permit a higher peak current and I^2t energy let-through than the current-limiting fuse, which can clear in a fraction of a cycle.

Table 16 shows an example of necessary application data for certain types of transfer switches suitable for use with fuses by specifying the circuit test data and fuse classes. Table 20 illustrates an example of application data by specifying the peak current and I^2t let-through limits of certain kinds of switches. Again, these characteristics will vary among different manufacturers. In one case the specific overcurrent protective devices are given along with the short-circuit test data, and in the other case the maximum limits of the switch ratings are given. In either case, the user can compare the switch capabilities with any type of fuse to be employed.

Figures 61 and 62 show the example system previously discussed with Figs 59 and 60 except that current-limiting fuses are used instead of circuit breakers. The transfer switch has a specific maximum allowable peak let-through current rating. Peak let-through curves based on available fault current and the clearing time of the fuses in the current-limiting range are available from fuse manufacturers. It would be an unusual case where selectivity in this range presents a problem in the coordination scheme.

6.3.8 Ground-Fault Protection. Although ground-fault protection of equipment is not normally required on the alternate source for emergency systems, ground-faults can occur on such systems and they can result in equipment burn-down. Because of the emergency nature of such systems, automatic disconnect in the event of a ground-fault is not required by code. However, detection of such a fault is desirable. It is good practice to provide an audible and visual signal device to

Fig 61
Emergency Power System with All-Fuse Protection

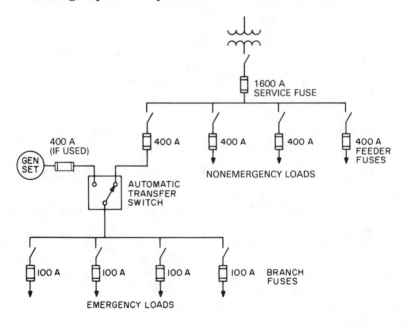

indicate a ground-fault condition. Instructions should also be provided to advise what course of action should be taken in the event of a ground-fault indication. Under certain conditions this is now a requirement of the NEC [12]. Additional consideration should be given to the possibility of a ground-fault on the load side of the transfer switch when energized from the normal source. For further discussion, see Chapter 7.

6.4 Generator Protection. When an emergency generator is running, it supplies the power for critical loads. In this role it is the most critical and vital element in the power supply. Manufacturers may not always provide all the basic protection needs for generators and also leave options to choice by the user. In any application, the user should evaluate his needs and the critical nature of the load before deciding on the best protection scheme. The protection scheme chosen should ensure reliability of this power source. Unlike applying standard practice in generator protection, a user should also rely on operating experience and sound judgment in determining whether or not standard protection schemes allow the required reliability of emergency power. For instance, whether or not an installation is

Fig 62
Coordination Chart of Emergency Power System with All-Fuse Protection

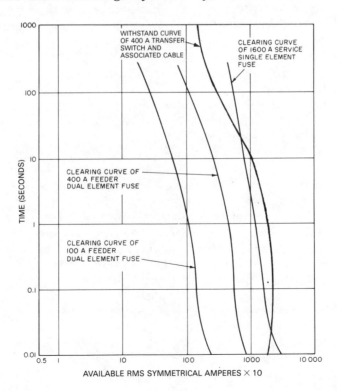

remotely or locally operated may determine if an audible or visual alarm will suffice in a condition where normally an automatic shutdown is used. Where power is extremely critical, it is wise to evaluate the consequences of loss of power to critical loads versus damage to the emergency power source. Caution should be exercised, however, when emphasizing reliability, to ensure that applicable codes and standards are not overlooked.

6.4.1 Codes and Standards. Some codes and standards that address emergency or standby generator protection are listed in 6.11 as: [6], [8], [10], [12], and [16].

6.4.2 Armature Winding Protection. Of most concern in winding protection is overcurrent, primarily from short circuits and overloads. The basic protective devices employed are circuit breakers or fuses. But the size of the generator set can drastically influence the method of use and trip levels for such devices. Also, overvoltage protection needs will differ between large and small machines (that is, high-voltage and low-voltage ratings).

6.4.2.1 Overcurrent Protection. The need for overcurrent protection for large machines is commonly taken for granted. For purposes of this chapter, these are machines that have enough mechanical and electrical stored energy, and thus available short-circuit current, to allow easy coordination between the generator and other branch circuit overcurrent protective devices. Overcurrent protection of the generator windings becomes important due to the cost of investment and also because protection will not seriously compromise reliability of the power supply to critical loads since overcurrent device selectivity is possible. Design of the protection scheme should follow standard practice, as a rule. Standard practice can vary from a single, simple molded-case circuit breaker to a larger power circuit breaker with complex relaying for trip initiation. ANSI/IEEE Std 242-1986 [8] provides a useful guideline for recommended generator protection.

Smaller generator sets pose a more perplexing problem regarding available short-circuit current. These are machines with relatively little mechanical and electrical stored energy and typically low system X/R ratios, and under bolted fault conditions can slow down very rapidly. This has been discussed, to some extent, in 6.2. Many users and specifiers of engine-generators often assume the need for a generator breaker for protection of the windings under short-circuit conditions. There are conditions where unregulated short-circuit current is capable of damaging a small generator. However, it is probable that the output of the generator, with a specified excitation system, will reduce to practically zero before any serious damage occurs. Even for separate excitation, the output can be specified to reduce to the full-load rating or less. This was illustrated in Fig 56. In this sense, it is inherently protected and an overcurrent device serves little purpose in protecting the windings against short-circuit current. The dangers of assuming inherent protection are the uncertain conditions of unbalanced faults. For instance, excitation could be supplied from an unfaulted phase and unwanted sustained fault current could occur. If inherent protection is desirable, the user should consider deliberately specifying this feature.

The initial per-unit fault current of a generator is $1/x''d$, where $x''d$ is the subtransient reactance of the generator and is determined from a short-circuit test of the generator. The initial short-circuit current induces very high currents in the field and damper windings of the rotor (8–10 times the no-load value is common). The generator losses under a fault condition, therefore, involve not just stator winding losses, but also field and damper losses, as well as extra stray load losses in both rotor and stator steel. It is possible for these losses to break couplings and tear loose generator mounts if the generator is not securely mounted. At the present time there are no reliable parameters available in the industry to determine the exact losses, but calculations based on the positive, negative, and zero sequence reactances and resistances will give approximate values. For a close analysis, these losses can be compared to the stored kinetic energy and the mechanical shaft energy input to determine the effect of a fault on generator speed.

In the discussions above for large and small generator sets, obvious consequences and effects of generator damage are assumed. In reality, the damaging effects of short-circuit current on a generator are difficult to determine. Even for a manufacturer, the actual limits of short-circuit current and duration are difficult to determine, since the tolerable damage can vary from slight reduction in insulation life to actual physical winding damage rendering the machine inoperative. For the mechanical strength of a generator ANSI/NEMA MG1-1978 [10] specifies that a generator be capable of withstanding without injury a three-phase fault at its terminals for 30 s when operating at rated kVA and power factor with fixed excitation for 5% overvoltage. The sustained current in this case is typically less than rated current, as evidenced in Figs 56(a) and (b), and hence does not test the windings for heating damage. Requirements are also given for unbalanced fault conditions.

Assuming that any generator has an accurate damage curve readily available for use in coordination of overcurrent devices is erroneous. Of additional concern, is the allowance of an extra 25°C rise in winding temperature for emergency or standby application of a generator, as designated in ANSI/NEMA MG1-1978 [10]. Often, the regulator/exciter system is the limiting factor in a damage curve. For instance, a common specification for a generator set is that it sustains 3 pu fault current for 10 s. This may actually be a limit for the regulator and exciter whereby the stator winding is capable of more. Further inquiry will reveal that rotor heating becomes a limiting feature during unbalanced faults. Rotor heating is caused by the negative sequence current. ANSI/NEMA MG1-1978 [10] requires that generators withstand negative sequence current only to an $I_2^2 t$ value of 40. During a line-to-line fault, which produces the highest negative current content, this translates to a total $I^2 t$ of 120. However, if the protection is set at $I_2^2 t = 120$, a separate device would be needed to trip the excitation in 10 s, if the excitation system is the limiting factor in the 90 $I^2 t$ calculation. Therefore, little is to be gained in coordinating capabilities by raising the overcurrent setting above the total system limit. Through consultation with a manufacturer, assuring that all conditions are understood, a reasonably accurate time–current limit can be estimated.

To avoid confusion and misconception about damage limit curves, a better descriptive might be *capability* limit. This is reasonable considering cases where it is difficult to determine limits other than from standards such as ANSI/NEMA MG1-1978 [10]. The manufacturer should be consulted in determining the capability of a generator since it should account for all components in identifying the weakest part. These components would include the stator, rotor, damper, and exciter windings. Semiconductors, as well as components of the regulator, should be included.

Thus, with fault-sustaining capability, the possibility exists for damage to the generator windings (and possibly to the prime mover) if the fault condition is allowed to exist too long. Figure 63 helps to illustrate this point. The figure shows an example of a generator capability curve based on 90 I^2t thermal limit calculated from 3 pu current at 10 s and its relation to the fault-current decrement curves for a generator with and without an excitation system designed to maintain fault current. The figure does not represent an actual case, but simply illustrates the maximum level at which the generator protective device must operate to provide minimum protection for the generator. The fault current available elsewhere

Fig 63
Generator Balanced and Unbalanced Fault Decrement Curves
Relative to Generator-Capability Curve

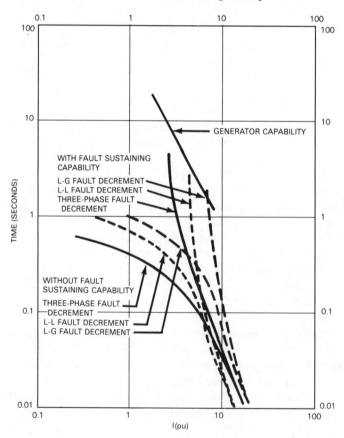

in the system would have to be calculated since the engine-generator manufacturer can only provide information on available fault current at the generator terminals. Another observation, not normally evident without the use of fault-current decrement curves, is that the sustained fault currents for line-to-line and line-to-ground faults are typically higher than the three-phase value. When only the three-phase fault-current decrement curve is available, the user should be aware that the overcurrent devices must trip faster for line-to-line and line-to-ground faults.

When a main breaker is to be used, a power breaker offers easy coordination with its adjustable tripping characteristics in long, short, and instantaneous time ranges. However, simple molded-case breakers cost less and are smaller in size for a given rating. But without adjustable tripping characteristics, they are more difficult to coordinate with other overcurrent devices or with the generator-capability limits. Some solid-state breaker-tripping devices offer constant I^2t characteristics that are suitable for generator protection. Relays offer either constant I^2t curves or extremely inverse curves which are a close approximation. Winding temperature detectors do not respond rapidly enough to provide this short-time thermal protection. Most fuses do not have a constant I^2t characteristic in the regions of the generator current, and so do not offer equal coordination with downstream breakers with variable generator fault currents as produced by different types of faults.

Proper selection and application of the overcurrent protective device will depend on the accuracy of the generator capability limit, available fault current, and the selected setting of the overcurrent device. Figure 64 shows an easy mistake to make in the application of a protective device. The generator in the example is rated 100 kW and 151 A at full load. A typical molded-case circuit breaker is applied at 125% of the generator rating, which assumes the load rating is equal to that of the generator. In this case, a 200 A breaker is used. The dotted portion of the decrement curve shows that it would take at least 80 s for the breaker to initiate a trip for 3 pu sustained current, if the breaker does not trip in the instantaneous region. Thus the generator would have to have a pu I^2t rating of well over 720 to be adequately protected. And, of course, the long-time trip region of the breaker is well beyond an I^2t rating of 90, which in this case is the assumed protection limit of the generator. The fault decrement curve (solid line) shows that the breaker's only chance of tripping is in the instantaneous region. Beyond 0.08 s the breaker will not see the fault current. However, since the current is much below the I^2t curve, generator protection should not be a concern.

In the case illustrated by Fig 64, if 3 pu fault current is provided by the generator, a better application of the same type breaker could only be achieved if less than full-load rating of the generator is applied allowing a lower breaker trip setting in both the long and instantaneous regions. This requires a generator sized for more than the allowable load. Otherwise, a breaker with better and more extensive trip adjustments, or a properly rated fuse, should be applied.

As pointed out earlier, damage to a generator because of lack of an overcurrent protective device in the generator output can seriously affect its availability to serve its intended purpose. The problem often facing designers is whether or not a protective device adds reliability or decreases reliability. This is not always an easy

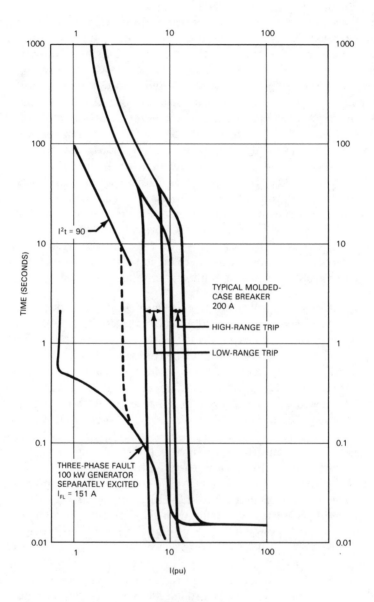

**Fig 64
Misapplication of Typical Molded-Case Circuit Breaker
with Limited Short-Circuit Current**

analysis. However, when an overcurrent protective device is required for generator protection and is properly applied, it will not operate unless damage to the generator is imminent.

Differential relaying should be considered to identify internal faults when two generators are operated in parallel. This allows identifying the faulted machine with the relaying to operate the proper breaker. Overcurrent devices in the generator output alone would sense equal fault currents and both generators could be disconnected on the occurrence of a fault in only one generator. When more than two generators are applied in parallel operation, overcurrent devices on the generators' output can function selectively for internal faults since each would sense more fault current than the others.

Automatic synchronizers will usually be employed for legally required systems or other systems where start-up time is critical. There should be an independent device to check the synchronizer's selected point of closure. The checking device should be of a quality that will not materially reduce system reliability. In addition, manual synchronizing should be available as a back-up, with a separate synchro-check device.

When the power system is grounded, the scheme employed warrants special attention. Multiple neutral ground connections can cause problems if ground current paths are not carefully designed. On large systems using special ground fault sensors for the generator protection, incomplete sensing or erroneous tripping can occur if the ground-fault current does not follow a predictable path. This problem is not so serious, as far as generator protection is concerned, on small systems that depend on a phase overcurrent device to operate on a ground fault. Likewise, with an ungrounded system ground currents are not a critical concern to generator overcurrent protection so long as proper detection is employed and single line-to-ground conditions are eliminated as soon as possible. Chapter 7 explores in detail various aspects of grounding and handling of the system neutral.

6.4.2.2 Overvoltage Protection. Overvoltage protection of the main generator windings is a necessity sometimes taken lightly or even overlooked. Emergency and standby power systems and their associated distribution systems are relatively small in overall power rating and number of components compared to base-loaded systems (*normal* power supplies and distribution systems) and also often located well within the overall distribution system. Consequently, they are not usually subject to the same number and severity of overvoltage problems. But, because emergency and standby power must be reliable, if effectively applied, overvoltage protection should always be included in the design and application process. Certainly it would be an unusual case if an emergency generator were physically connected to an electric utility system such that it was subject to the same dangers of lightning transients. But other types of transient voltages from switching and restriking faults, for example, can be damaging and should be of concern.

Overvoltage protection is more critical for large generators with high voltage ratings than for smaller ones with low voltage ratings (600 V). Protection of the investment is one reason, but another is the decreased margin of safety in the insulation as voltage ratings increase. That is, the ratio of insulating capacity to system voltage is normally higher for smaller units than for larger, higher voltage-rated units. ANSI/NEMA MG1-1978 [10], for instance, stipulates a suitable high potential test for armature windings of 1000 V plus twice the rated voltage for most common types of ac generators. The test voltage for a 4160 V generator at this condition would be 12% higher than twice the rating. The test voltage for a 480 V generator would be about 100% higher than twice the rating.

Overvoltage conditions can be either of a transient nature or steady-state nature. The former is considered higher in magnitude and frequency, but of considerably less duration than the latter. Protection from each type of overvoltage condition is handled differently, but for the same reason: to preserve the integrity of the insulation. Although difficult to measure or calculate, any overvoltage condition, no matter how short the duration, will stress the insulation and shorten its life. So this is always a concern.

Protection against transient voltages is most commonly provided by surge arresters in combination with capacitors. But unless the transient is high, as it would be from a lightning strike, surge arresters are not always effective, although capacitors will attenuate a surge at any level. Switching surges would rarely reach the spark-over value of a lightning surge arrester. When an emergency or standby generator is situated to be subject to lightning-induced transients, surge arresters should be applied preferably at the generator terminals and, if necessary, at the boundaries of the exposed part of the system, such as at the ends of the exposed portion of an overhead line. The exact clamping level of a surge arrester is not easily predicted. The margin of safety is greatly influenced by the expected frequency and magnitude of the surges, something a designer would determine largely by judgment. Each occurrence of an overvoltage will shorten the insulation life of the windings with an inverse relation between the magnitude and allowable duration. The manufacturer can provide the necessary information on the withstand capability of the winding insulation.

This will aid in determining the upper limits of allowable overvoltage and duration. However, some additional margin of safety should be considered. ANSI/IEEE Std 141-1986 [6] suggests limiting the magnitude to approximately four fifths of the certified test voltage. The permissible rate of rise of the transient is suggested as no more than the 60 Hz high-potential test peak value at a uniform rise time of 10 μs. This rate of rise is controlled by the application of appropriate surge capacitors at the generator terminals.

Other types of transients, common to industrial and commercial systems, result from switching actions that force current to zero which in turn, generate a transient voltage from the rapid collapse of the magnetic field. This type of overvoltage can result from the operation of devices ranging from circuit breakers to SCRs. Even the opening of a circuit by a conductor burning itself free from a fault can produce a transient overvoltage. In ungrounded systems, the neutral is free to change its electrical reference to ground under certain conditions, which can compound the transient effects of switching.

The most effective method of protecting an emergency generator from switching transients is careful design of the installation. Switching transients, as mentioned above, are rarely as severe as those from lightning and are not normally given the same consideration. Even though their occurrence can be more frequent than those from lightning, there is also more inherent protection within a distribution system, such as the number of cable circuits and connected equipment that act to greatly reduce the magnitude and slope of surge voltages. Grounding of a system neutral minimizes additional voltage-to-ground stress caused by switching surges and ground faults (especially restriking ground faults). Additionally, regardless of whether a system neutral is grounded or not, switching can cause overvoltage stress between windings from the oscillatory effect of voltage when a switch clears a fault.

ANSI/IEEE Std 141-1986 [6] provides a detailed analysis of the causes and effects of surge voltages along with recommended application practices that afford protection.

Steady-state overvoltage conditions are not as serious as transient overvoltages because of their lesser magnitude and because they are easier to control. They are of concern, however, because of their duration, and again it is emphasized, insulation life is shortened by these overvoltages. Some common causes are overexcitation, accidental contact with a higher voltage source (this can sometimes be of a transient nature), and shifting of the system neutral in ungrounded systems.

Overexcitation can result from unlimited control of an exciter and voltage regulator or excessive leading reactive power from external devices like capacitors that are designed only for use with the normal power supply. Accidental contact with a higher voltage source happens in rare cases. This is usually the result of insulation breakdown between a high and low winding in a transformer. In an ungrounded system, the neutral would be raised above ground reference by an amount equal to the high voltage and thus raise each phase above ground accordingly. The occurrence of this condition is so rare that deliberate measures to guard against it are rarely taken. Proper enclosures and shielding of the conductor systems provide the proper safeguard. In a grounded system, the ground of the secondary and primary (presuming the high-voltage side is physically referenced to ground) will complete a short-circuit path when the two voltage sources come in contact, allowing overcurrent devices to operate. An initial voltage transient is not prevented but the duration is minimized.

Shifting of a system neutral and corresponding phase overvoltage to ground under steady-state conditions in an ungrounded system is caused from single line-to-ground faults and unbalanced loading conditions. Both occurrences can be singularly classified as unbalanced conditions. If an ungrounded system is necessary, then balanced loading and ground detection serve to protect a generator from the deteriorating effects of overvoltage, but only if prompt action is taken to correct the unbalance when it occurs. Specifications for the generator winding should assure line-to-line voltage rating of the line-to-neutral winding in a wye-connected generator to be applied in an ungrounded system.

6.4.2.3 Harmonics. The increased use of static power converters has brought with it an increased concern of the effects on emergency and standby power generators. Some of the most serious effects pertain to proper operation and quality of power. But, there is also cause for concern for generator protection. Some obvious concerns are listed:

(1) High-frequency harmonics cause additional losses and heating in a generator. A significant portion of the losses occur in the rotor where heating is not measured.

(2) Overcurrent relays can become more sensitive because of the skin effect in the current-sensing elements. Electromechanical and static relays may have altered performances, with both the duration and amount of change following no predictable criteria. The overall result is that a relay may not operate as accurately as specifications show.

(3) Although little quantitative information exists on the effect of harmonics on circuit breakers and fuses, it is expected that their current-carrying capacity would be reduced, as heating of thermal elements is increased due to skin effect.

6.4.3 Rotor and Excitation System. A field circuit breaker is a positive means of protecting the rotor and excitation system from damaging overcurrents due to external faults, underspeed, certain load conditions, or component failure within the excitation system. A field breaker, however, is not considered adequate protection by itself for the stator windings and should not replace an overcurrent device in the generator output, if one is required.

Solid-state exciter/regulator systems are commonly available and are becoming more and more versatile. Common protection features include over and under excitation limiting, automatic voltage sensing between manual and automatic regulation circuits to allow smooth transfer from automatic to manual operation and vice versa, and failed rectifier detection.

In the brush-type machines (including those with static exciters) the field breaker may be located in the leads to the slip rings or in the field leads of a self-excited exciter or the excitation supply of a separately excited machine to reduce breaker size and cost. On brushless generators only the last two locations are available, since the main rotor leads are not accessible. Location in the regulator power input is usually preferred since it provides short-circuit protection to the regulator and the regulator performs the field discharge function. In this position, the breaker may contain additional contacts that provide short-circuit protection for the voltage-sensing circuitry. Thus, excitation would be removed to prevent over-excitation when sensing is lost due to a fault or protective device operation.

6.5 Prime Mover Protection

6.5.1 General Requirements. The most direct form of overload protection, still maintaining some degree of reliability, would be load shedding. Depending on the severity of stability problems, breaker position or frequency sensing might be used to initiate action. Instantaneous automatic load shedding where multiple generator sets are used, when one or more generators are lost, would assure available power for the remaining, more critical loads. On smaller systems, especially where only one generator is used, frequency sensing to shed load might provide a more reliable power supply if upset conditions do not always require load shedding. A combination of the two methods would allow a system to instantaneously shed selected loads, with frequency sensing employed as backup to shed additional load as necessary. It is common practice in some cases to employ underfrequency relays as secondary sensing devices to trip selected load breakers in multiple steps, with a time delay between each step. A stability study would determine frequency settings at each successive step. The study would also determine how fast load shedding should occur and thus determine the type and speed of equipment to be used. The scheme would normally be designed to maintain enough generation to prevent total blackout regardless of the load conditions.

Overload protection for a prime mover can also be aided by assuring that frequency sensitive voltage regulation is used. Maintaining a constant voltage-frequency ratio minimizes effects of overload and allows a unit to more easily regain normal voltage and frequency after an overload. The output voltage of the generating set would decrease in proportion to the frequency (prime mover speed). The use of a nonfrequency sensitive voltage reference could require, in some cases, a load reduction of from 50–60% to allow return to rated speed. Although some manufacturers can provide either type of voltage reference, it may be up to the user to specify his preference. Application of frequency sensitive voltage regulation, however, should not overshadow the importance of proper matching of generator or prime mover torque characteristics.

Reverse power relaying is an important form of protection for prime movers. It will prevent motoring when generator sets are operating in parallel and in another application prevents overload of a generator set by fast relaying when power flow is into an electric utility system. In prevention of generator motoring, the user should be aware that some prime movers are less susceptible to damage than others. Sensitivity is more critical, for instance, on turbines than reciprocating engines, and nuisance tripping can occur if this is not accounted for.

Protection against motoring of a prime mover guards against overheating or cavitation of blades on a turbine and possible fire or explosion from unburned fuel in a reciprocating engine. A relay to detect reverse power flow would normally be applied as backup protection to mechanical devices designed to detect these conditions. A time delay can be applied to prevent nuisance tripping on momentary reverse power surges such as may occur during synchronizing. Some typical values are listed below showing reverse power required to motor a generator when a prime mover is being spun at synchronous speed and at no input power:

Condensing turbine	3% of nameplate, in kilowatts
Noncondensing turbine	3% of nameplate, in kilowatts
Diesel engine	25% of nameplate, in kilowatts
Hydraulic turbine	0.2 to 2.0% of nameplate, in kilowatts

6.5.2 Equipment Malfunction Protection. Standard protective devices and numerous options are provided with prime movers by manufacturers. The equipment investment and nature of critical loads determine how to apply this protection. Some shutdown devices might be considered for alarm only, such as where installations are attended, and a malfunction can be quickly investigated. When it is determined that a shutdown is necessary for a malfunction condition, protection integrity can be maintained and power supply reliability enhanced if the malfunction is such that an alarm and subsequent shutdown can be employed.

High water temperature, high oil temperature, low oil pressure, overspeed, high exhaust temperature, and high vibration are typical examples of malfunctions that lend themselves to two-level protection as suggested above. On large machines, this is usually an insignificant investment to make. It is common practice in many installations and often required by machine manufacturers to also provide meters for continuous monitoring of the above parameters by personnel on site or at remote locations. When remote controlled machines are located in areas not easily accessible, readouts of individual machine parameters become more significant. Cases have occurred where engines and turbines have been damaged because operating personnel with remote control capability restarted the units or continued operation of the units under malfunction conditions without knowing specifically which malfunctions existed.

Large generator sets requiring sophisticated and complex control systems often utilize uninterruptible power supply (UPS) systems for control power. Protection of power supplies of this nature is vital to the total machine protection.

Reciprocating engines and turbines require different philosophies for protection by nature of their design and operation. Smaller reciprocating engines typically have protection furnished for high water temperature, low oil pressure, overspeed, and failure to start. Larger units might also include high oil temperature, high vibration, antimotoring, and protection for sophisticated control and excitation systems.

Protection for combustion turbines would add such protective devices as failure to light, failure to reach self-sustaining speed, and exhaust temperature limits and control. Overspeed and vibration are obviously more critical on turbines due to high-speed operation.

Electric motor-driven auxiliary equipment on large generator sets, such as lube oil pumps and cooling water pumps, are vital to the protection and reliability of the generator set. Reliability might be increased and protection maintained in locally operated critical installations by removing thermal overload trips from motor-control circuits and alarming only on an auxiliary motor overload. In this instance, lube oil temperature and levels, cooling water temperature, and auxiliary motor loading should be closely monitored to protect equipment investment.

6.5.3 Fuel System Protection. The importance of protection for fuel systems needs very little explanation regarding the reliability of the power supply. For instance, it is obvious that low pressure and level alarms can prevent needless shutdown or failures of emergency generator sets requiring a high degree of reliability. Building codes and fire insurance regulations aid in determining optimum locations of self-contained fuel systems. Refer to Chapter 4 for storage recommendations and ANSI/NFPA 30-1984 [11] for requirements on protection for fuel piping.

It should be noted that gasoline and diesel fuels can deteriorate if left standing unused for long periods of time. Some provision should be made for periodically burning or replacing the fuel.

6.6 Electric Utility Power Supply. An electric utility supply serving in a *standby* role is typically protected no differently than if it were the *normal* supply, with one exception: the problem of possibly connecting two sources out of phase must be dealt with. A standby electric utility supply is either disconnected from the other source and closed in when needed, or it is in parallel and synchronism with the other source and supplies additional power when needed. In the latter case, the available reserve power can be considered a standby source, but is more commonly considered part of a redundant normal source. In this sense, it is different from what is usually considered standby.

Connecting two sources out of phase is the greatest danger when an electric utility source is used as a standby source because of the switching or transfer requirements. If the standby source is normally disconnected until needed, it can be *out-of-sync* with the load residual motor voltage when closed in. In this case, the transfer controls should include either in-phase monitoring for fast bus transfer or, if the load can tolerate an outage possibly as long as a few seconds, a time delay to assure the residual motor voltage has decayed to a safe value before switching to a standby source. Refer to Fig 19 in Chapter 4 for a guide to typical motor voltage decay curves. It should be remembered that power factor capacitors will increase the open-circuit time constant of the motors.

Switching from a standby power supply back to the normal supply would demand the same synchronizing requirements whether the standby source is an electric utility or an in-plant generator. In either case, closed or open transition can be employed depending on load requirements, and the issue of synchronizing would be the same.

Because of the exposure of an electric utility line to uncontrollable disturbances, this source is rarely employed as an *emergency* source, as defined in this standard. Therefore, special techniques and protective equipment to preserve reliability, such as those recommended for other in-plant emergency or standby sources, are not considered necessary or practical. Standard guidelines for protective equipment for electric utility ties are covered quite well in ANSI/IEEE Std 242-1986 [8]. It will suffice to say here that close cooperation between engineering representatives of the electric utility and user is necessary to assure the needs of each are clearly defined and understood. An accurate description of the load and its effects on the utility supply is essential in establishing proper protection of the supply while maximizing reliability. One additional important concern is that of ownership of the service equipment. If the user requires protection and control equipment not normally employed by the utility, he may choose to furnish and own the substation equipment to maintain this control.

Protective schemes vary in complexity depending on the size, economic investment of the system, and consequences of power loss, but the objectives are the same. The utility supply must be protected from the effects of faults and abnormalities in the user's system and, conversely, the user's system must be protected against the possible adverse conditions in the electric utility system.

Protection against ground faults raises special issues. Ground-fault sensing can be a problem if not carefully planned; see Chapter 7 for methods and considerations in handling ground faults.

6.7 Uninterruptible Power Supply (UPS). The basic components of a static UPS are the battery, rectifier or battery charger, inverter, and often a static transfer switch. Technically, a motor-generator set is also a UPS system, but is often referred to as a mechanical stored energy or rotary system. The discussion here is limited to the static-type system.

6.7.1 Battery Protection. Batteries provide the source of emergency or standby power in the application of a UPS system. The reliability of the system will heavily depend on the protection provided for the battery. Much of the protection provided is inherent in good maintenance practice as discussed in Chapter 8, but the importance of this subject warrants further discussion in this section.

Lead-acid batteries are especially susceptible to adverse effects from both being undercharged and overcharged. Continual undercharging promotes lead sulfate buildup which reduces battery capacity. Overcharging causes gassing of vented cells and heating of sealed cells. A lead-antimony battery experiences objectionable water consumption if a float voltage is too high and thus reduced battery life. Lead-calcium batteries are less susceptible to high water consumption.

Most battery specifications and recommendations are based on an ambient temperature of 77 °F (25 °C). Float charging or equalize charging a lead-acid battery in an ambient temperature higher than this *optimum* value can reduce battery life below its designed life if the voltage is not reduced accordingly. Typical recommended float voltages for fully charged lead-acid batteries are shown in the following list. However, in any specific installation a manufacturer's recommended values are the preferred choice.

Battery	Volts per Cell
Lead-Antimony	2.15–2.17
Plante	2.17–2.19
Lead-Calcium	
1.215 sp gr	2.17–2.25
1.250 sp gr	2.23–2.33
1.300 sp gr	2.28–2.37

When equalize charging by constant voltage is necessary to bring all cells to an acceptable charge state, a voltage level higher than a normal float voltage is used (provided the voltage level does not exceed the maximum tolerable level for connected loads). Equalize charging has a protection benefit for lead-antimony batteries because it counteracts the degradation of natural transfer of antimony from the positive to negative plate in an undercharged battery. An equalize charge maintains the integrity of a battery, but should not be applied longer than necessary. The worst condition for hydrogen evolution is forcing maximum current into a fully charged battery. Some recommended voltages and time periods for equalizing lead-acid batteries are shown in the following tables. These values are based on cell temperatures in the range of 70 °F (21 °C) to 90 °F (32 °C). As with the recommended values of float voltages in the table above, manufacturers should be consulted for specific or unusual installations and especially for charging requirements for other temperatures. Additionally, charging equipment should be evaluated as to whether or not it is ambient temperature compensated, thus accounting for different cell temperatures.

Lead-Antimony and Plante

Volts per Cell	Time (hours)
2.24	80
2.27	60
2.30	48
2.33	36
2.36	30
2.39	24

Lead-Calcium

Volts per Cell	Time (hours)		
	1.215 sp gr	1.250 sp gr	1.300 sp gr
2.24	222	—	—
2.27	166	—	—
2.30	105	—	—
2.33	74	166	—
2.36	50	118	200
2.39	34	80	134
2.42	—	54	91
2.45	—	36	62
2.48	—	—	42

The initial charging of a lead-acid battery after shipment or installation is very much like equalize charging after a discharge. The difference is that the time period is longer — typically twice as long.

An important protection feature for a lead-calcium battery is correct sizing of the ampere-hour capacity. A lead-calcium battery cannot take deep discharges, so control of the discharge end voltage is more critical than for other battery types.

Nickel-cadmium (NICAD) batteries, because of their construction and chemical composition are more tolerable to high charge rates and high ambient temperatures than lead-acid batteries. However, overcharging (continued charge current after full charge reached) still causes oxygen and hydrogen evolution. A sealed cell would experience pressure buildup and temperature rise.

A vented cell would experience gassing and loss of water from the electrolyte. Approximate charge voltages per cell related to battery temperature are shown in Fig 65 for vented-cell NICAD batteries.

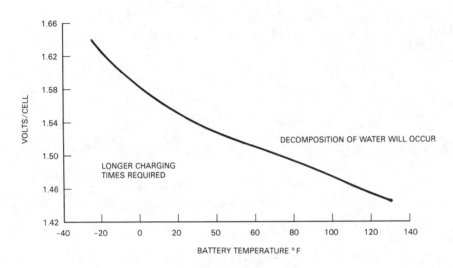

Fig 65
Charge Voltage versus Temperature for Vented Cell NICAD Batteries

The detrimental effects of overcharging batteries are compounded with high ambient temperatures that add to the increased temperature resulting from charging. High temperature increases water consumption by evaporation, and the resulting increased charging current from high temperature causes more water consumption through oxygen and hydrogen evolution. Acid fumes from lead-acid batteries corrode terminals (and other equipment in close proximity). Hydrogen evolution can create an explosive atmosphere if the area is not adequately ventilated. Sealed cells can build up dangerous pressures if temperature is too high. For these reasons, controlling ambient temperature may be the single best protection for preserving battery integrity and maximizing life.

When the environmental conditions, such as ambient temperature, cannot be controlled adequately, chargers should be equipped with the ability to automatically compensate for temperature changes by reducing charge potential with increasing temperature. Some battery manufacturers incorporate devices within the cells of the battery to protect against the effects of overcharging and overdischarging. For instance, semiconductor devices can be used to shunt part of the charging current if voltage and temperature become uncoordinated or they can prevent cell polarity reversal on excessive discharge.

Good practice would be to maintain the battery environment between 60 °F (16 °C) and 90 °F (32 °C) with the optimum temperature at or near 77 °F (25 °C). The battery area should be adequately ventilated to prevent hydrogen buildup with exhaust to the outside atmosphere. Under normal conditions, the lower explosive limit of hydrogen mixed with air is 4%, so ventilation should be adequate to ensure a mixture well below this level. Internal gases in the cell can also ignite from an external ignition source. If this can be a problem, consideration should be given to installing *explosion-resistant vents* to prevent this ignition. Cells should be located so that an even distribution of temperature between cells is possible. Outside devices that might cause concentrated sources of heat to some cells should be avoided in the location of the battery.

An area of battery protection often overlooked is overcurrent protection. It is often impractical to apply and install overcurrent devices, such as breakers or fuses, to protect a battery and its cable from short-circuit currents because of the physical layout of the equipment. However, a designer should be aware that without overcurrent protection a battery can be damaged and, in some cases, the battery cable may inadvertently be the only fuse protection available. A battery sized for 50 A for 2 h might be capable of 2000 A of short-circuit current. The short-circuit current available from a battery is the voltage rating of the battery divided by the resistance between the battery terminals; more simply, it is the cell voltage divided by the internal resistance of the cell as expressed by the formula

$$I_{sc} = \frac{\text{voltage per cell}}{\text{internal resistance per cell}}$$

The internal resistance of a cell is not normally published in most literature available from manufacturers. If this is the case, consult manufacturers for this information. A useful guide in estimating the internal resistance of a battery or cell is by the use of the two rated end voltages along with the rated current at each of these voltages. For best results the time period chosen to obtain these voltage and current values should be short, typically 1 min or less. With this information, the internal resistance R between the points of voltage measurement (such as individual cell or the whole battery) is simply

$$R = \frac{\Delta E \ (1 \text{ min})}{\Delta I \ (1 \text{ min})}$$

or

$$R = \frac{E1 - E2}{I2 - I1}$$

As an example, assume a 200 ampere-hour lead-acid battery is being considered. Specifications show 382 A for 1 min to an end voltage of 1.75 V/cell and 748 A for 1 min to an end voltage of 1.50 V/cell. The cell internal resistance R is estimated by

$$R = \frac{1.75 - 1.50}{748 - 382} = 0.000683 \ \Omega$$

In this example, the short circuit, I_{sc}, available at a nominal 2.17 V/cell is

$$I_{sc} = \frac{2.17}{0.000683} = 3178 \text{ A}$$

Loads supplied by UPS systems are critical and low-voltage alarms are necessary to alert personnel when the power supply integrity is in danger, so this protection feature is recommended with any type of battery system. However, a lead-acid battery is especially susceptible to damage if excessively discharged, so a low-voltage alarm serves to protect the battery as well as the load.

6.7.2 Battery Charger Protection. Overcurrent protection for a battery charger, like most devices, should include both overload and short-circuit conditions. In the case of a charger, the current required in charging a severely discharged battery can easily cause overload conditions aside from the inverter load. Simple overload protection should allow for the charging requirements of a battery.

To charge a battery that has been discharged to any large extent requires that the charger output current be regulated to stay within acceptable limits. This protects both the charger and the battery. Of particular concern in a charger is the rectifier circuit, especially if it is connected directly to the output. Current ratings should be known in order to assure good protection.

Most chargers are provided with a current-limiting feature for protection. Not all chargers have the same current-limiting characteristics. A user should evaluate the particular charger under consideration and the short-circuit characteristics of its output to assure proper overcurrent protection. Some chargers have current-limiting features designed primarily for protection in charging a discharged battery. Other simple overload conditions would also be protected against. However, the charger may not be able to withstand a bolted fault on its output. The fault current at zero voltage could easily be higher than the charger's current-limit value. In this case, additional output protection is needed, such as breakers or fuses, unless the charger is equipped to automatically shut down when a dead short occurs. Some charger specifications include current limiting designed to allow withstand of a dead short on the output, even continually. In this case, an output breaker or fuse adds flexibility but is only redundant for short-circuit protection.

Regardless of the reason for a condition of current limiting, if it is expected to occur frequently or for long durations, consideration should be given to adding thermal protection in addition to the current limiting feature. Some chargers are equipped with a thermal circuit that automatically reduces output current if a predetermined temperature is exceeded.

If heavy inrush currents are possible (loads other than battery-charging current), a problem with the charger going into current limiting or tripping of the output protective device can occur. This can affect the applied load by dropping all or part of it off-line. Some chargers can be equipped to allow gradual load application to prevent nuisance tripping or current limiting.

A protective device on the ac input to the charger is also necessary to protect the charger from extensive damage from internal faults. Manufacturers provide charger input ratings making protective device ratings easy to determine.

Optional protective devices that would usually be supplied with a battery charger, although not necessarily part of the protection scheme for the charger alone, should also be mentioned here. Ground detection with light indication or voltmeter readout can be supplied for ungrounded dc systems. High and low dc voltage relays serve to protect the load and charger. High-voltage detection serves to protect semiconductors, since high-voltage increases leakage current which increases temperature. On some chargers an ac power failure disconnect relay can be furnished to open the input circuit of the charger and thus protect the battery from unnecessary discharge back through the charger.

6.7.3 Inverter Protection. Protecting an inverter from overcurrent conditions and still maintaining a reliable power supply to critical loads takes careful planning and evaluation of the load and power supply requirements. An inverter is considered a *soft* source in that it normally does not have abundant short-circuit current capability. Protecting the output allows very little tolerance between the inverter rating and protective-device rating, but at the same time normal inrush currents must be tolerated.

Like battery chargers, inverters also are commonly supplied with current-limiting capability for protection. This adds to the difficulty in sizing branch circuit overcurrent devices because they must operate fast enough on branch circuit faults to prevent current-limiting action by the inverter and still allow enough flexibility for inrush currents. Typical ratings of an inverter might be 125% of full load for 10 min and 150% of full load for 10 s. Often this is sufficient for selective coordination of branch circuit devices if the load is distributed among a sufficient number of branch circuits. Still, high-speed fault clearing is likely necessary to prevent inverter current limiting. For example, the inverter would start current-limiting action at 150% of its rating (some inverters would start limiting current before this value is reached), so branch circuit overcurrent devices must be set well below this value. This could pose a problem even for fuses if they are designed for the inrush requirements of some inductive loads.

One solution to this problem of preserving the integrity of the power supply to the load is a design allowing the static transfer switch to switch to the alternate ac supply on fault or heavy inrush current conditions. The alternate supply would have the short-circuit capability to easily allow clearing of the fault or the inrush current requirements and then automatic retransfer to the inverter would take place. This alleviates the problem of precise coordination of the branch circuit overcurrent device with the current-limiting action of the inverter. It should be understood, however, that if the alternate ac source is not available at the time of necessary transfer, the inverter will go into current limiting, if it has this feature, and all loads will be affected if the faulted branch circuit does not clear. If this UPS is designed for total dependency on transfer to the alternate source for branch circuit overcurrent conditions (that is, current-limiting feature not provided), the inverter will shut down, trip off, or be damaged and, again, all of the load will be affected.

The reliability required of the power supply for the load will determine how the distribution system is designed and the type of protection provided for the UPS. There are various ways to protect the inverter from overcurrent, but in each, reliability is affected to some extent. High-speed semiconductor fuses can be applied at the inverter output, but this makes coordination with branch circuit overcurrent devices difficult, if not impossible. These fuses could be applied to the branch circuits to make fast clearing of branch circuit faults possible. The main disadvantage here is the high cost of these fuses. As mentioned above, available short-circuit capacity can sometimes be a problem. An oversized inverter would help to alleviate this problem but, again, investment cost would be high. Applying a transfer scheme to automatically transfer high inrush currents and short-circuit currents to an alternate power supply offers a relatively simple approach to over-current protection. This makes selective coordination of overcurrent devices much easier. This protects the inverter and, in short-circuit conditions, allows isolation of the faulted circuit, thus minimizing the disturbance.

Even though inverters are provided with current-limiting action, they are not always rated to tolerate short-circuit conditions. To protect against the event of an alternate ac source not being available to handle short-circuit current conditions, an overcurrent device should be provided in the inverter output. Since fuses are generally faster than breakers, they should be given first consideration. However, since semiconductor devices are involved, even a fuse cannot guarantee preventing all damage, but it can assure minimizing the damage. Another overcurrent condition that should be protected against on the output is out-of-phase switching.

Another common problem in the application of UPS systems is inadequate venti-lation for the inverter. Inverters generate a considerable amount of heat and some means of removing this heat should be provided. Good ventilation is the best method of preventing overheating. The installation design should assure that this ventilation is not blocked, and personnel should be made aware of the danger to the inverter if ventilation is blocked. Each inverter has a maximum operating temperature rating, and exceeding this temperature should be prevented. Inverters can also overheat under prolonged short-circuit conditions even though they may be rated to withstand a continuous short circuit on their output. Thermal cutouts can be provided to protect against this heating effect. Rare cases have even been reported where overheating occurred at no-load conditions.

Input overcurrent devices for inverters will minimize damage caused by internal faults. Again, fuses should be given first consideration for high-speed clearing. Another source of high current in the input protected against by fuses or breakers is an accidental reverse polarity connection. An important option offered for most inverters is a low-input voltage alarm. Since the input is connected to batteries in a UPS and to the output of a battery charger, the possibility always exists for low-input voltage. The inverter can only maintain a specified voltage at its output if the input is within specified limits. Strictly speaking, this feature provides protection for the battery against excessive discharge, but it also provides protection for the load against the adverse effects of low voltage, especially if transfer to an alternate ac source is initiated when the low-voltage limit is reached.

6.7.4 Static Transfer Switch Protection. Protecting a static transfer switch from excessive overcurrent requires coordination of the switch rating with the available fault current from the ac supply with the highest available fault current. Static transfer switches commonly have a short-circuit rating at least 1000% of the continuous rating for 1–5 cycles. This is more than adequate for the fault-current availability of an inverter, but careful overcurrent device setting will be required when the fault current is from a larger alternate source. Again, since semiconductor devices are involved, fast fuse operation is necessary for adequate protection. Another feature that adds protection to the switch is fast removal of gating voltage, which effectively opens the switch. Most commonly, switches are protected by applying the overcurrent device on the input side. The manufacturer or UPS supplier should be consulted for static transfer switch withstand ratings for proper application of overcurrent protection.

Even though the inverter source is not likely to supply a lot of short-circuit current, especially if it goes into current limiting on a short circuit, an inverter output protective device serves also to protect the static switch. This can be the case when a fault occurs downstream of the switch (or an overload) and the alternate source is unavailable for transfer.

6.7.5 Overvoltage Protection. As previously stated, the basic components of a UPS are the battery, a rectifier or battery charger, and often a static transfer switch. These devices contain semiconductors making protection uniquely different from that of more tolerant equipment, such as motors, generators, transformers, electromechanical transfer switches, etc. In dealing with semiconductors, surge protection becomes a critical application criteria. Simply increasing insulation thickness and dielectric strength is not feasible because of the necessary properties of the devices, so effective protection must be applied externally. Sources of surges and transients are many, and without proper protection semiconductors can be easily damaged. Even if the damage is not visible, they can be stressed such that their life is drastically shortened. Even a relay coil can produce a surge in the order of thousands of volts under fast-contact current-interrupting action if it is not itself equipped with surge suppression such as a diode or varistor to handle the steep wavefront and high-frequency surge associated with the interruption.

Equipment manufacturers are usually best suited to determine proper surge protection for the semiconductors in their equipment. However, they cannot always be accurate in their assessment of the application of the equipment. Even though surge protection may be standard with some manufacturers, the source and magnitude of possible voltage transients and application conditions are often underestimated. It is not uncommon for a surge suppressor to be damaged if operating conditions allow the suppressor to be subjected to steady-state voltages exceeding its rating. If the user supplies this type of protection, he must know the dielectric strength of the semiconductors to be protected to determine the rating of the surge suppressor.

There are many types of surge suppressors available with different operating characteristics. Some are fast acting but may allow a higher let-through voltage than another that is slower acting but capable of lower threshold settings. Some of the more expensive surge suppressor packages employ a combination of both characteristics for improved protection.

Isolating transformers provide some measure of surge protection because of the physical isolation effect on circuits. But because of their magnetic coupling and distributed capacitance between windings, they are not perfect isolators and often additional surge protection is required. So-called *line conditioners* offer the best protection in that they provide line isolation, surge protection, high-speed voltage regulation, and in some cases a degree of wave shape filtering for protection against harmonic distortion, all in an integrated package.

Methods used in grounding equipment cabinets affect the magnitude of insulation stress to ground especially when equipment is exposed to high transients typical of lightning strikes. The inductive voltages (due to inductance of grounding conductors) can stress equipment insulation. Shortening ground conductor lengths or paralleling ground conductors, or both, are effective methods in reducing the inductance of ground circuits. Each installation will be unique in proper procedures for adequate protection and personnel safety.

6.8 Equipment Physical Protection. Engine-generator sets and fuel supplies and their associated equipment require careful planning in application to prevent damage from physical abuse and environmental conditions. Radiator cooling on engines should be given special consideration, when possible, since a self-contained cooling system does not require external piping connections that could be subject to damage. Some climates, such as where dust is prevalent, require that air inlet filters be of special design.

Most other devices and equipment discussed previously, when applied in emergency and standby use, are usually located where only qualified personnel have access. When this is not possible, enclosures should be designed to prevent easy access. Provisions for locking enclosures with hinged panels and doors can easily be supplied by most manufacturers. Manufacturers should also be thoroughly knowledgeable of the location and environmental conditions for application of proper enclosures, materials, and seismic protection. For example, corrosive or humid atmospheres may require heaters or special coating or materials to protect exposed metal parts or circuits. Locations in proximity to large reciprocating machinery might require special seismic protection. Fuel systems naturally require special consideration in protection from leaks, contaminants, etc.

Manufacturers can usually provide suitable racks for batteries that provide protection and easy maintenance. It is important to have properly sized racks that space cells to minimize overheating and corrosion problems, allow easy maintenance, and still utilize available space in an optimum manner. Connections between cells should minimize strain on battery posts. Manufacturers can also assist in designing installations for extreme seismic conditions, such as where earthquakes are a problem. Additional information on installation and environmental conditions can be found in Chapter 5 of ANSI/NFPA 110-1985 [13].

6.9 Grounding. Reliable ground fault protection schemes require a careful analysis of the system- and equipment-grounding arrangements. Chapter 7 covers the basic functions and requirements for system and equipment grounding. Several grounding arrangements that are applicable to emergency and standby power systems are illustrated in Chapter 7.

6.10 Conclusions. Protection of emergency and standby power systems determines to a great degree the reliability of the systems. Although the total system should be considered in the protection scheme, overall system protection and reliability is only as good as that of the individual components. With emphasis on protecting power supply investment, a designer should be sure that the total investment is not overly compromised. The designer should carefully evaluate standard and optional protection schemes supplied with equipment and apply the design best suited to his needs for equipment protection and power supply reliability.

6.11 References. This standard shall be used in conjunction with the following publications:

[1] ANSI C37.16-1980, American National Standard Preferred Ratings, Related Requirements, and Application Recommendations for Low-Voltage Power Circuit Breakers and AC Power Circuit Protectors.

[2] ANSI/IEEE C37.13-1981, IEEE Standard for Low-Voltage AC Power Circuit Breakers Used in Enclosures.

[3] ANSI/IEEE C37.26-1972 (R 1976), IEEE Standard Guide for Methods of Power-Factor Measurement for Low-Voltage Inductive Test Circuits.

[4] ANSI/IEEE C37.90-1978 (R 1982), IEEE Standard Relays and Relay Systems Associated with Electric Power Apparatus.

[5] ANSI/IEEE C37.95-1973 (R 1980), IEEE Guide for Protective Relaying of Utility-Consumer Interconnections.

[6] ANSI/IEEE Std 141-1986, IEEE Recommended Practice for Electric Power Distribution for Industrial Plants.

[7] ANSI/IEEE Std 142-1982, IEEE Recommended Practice for Grounding of Industrial and Commercial Power Systems.

[8] ANSI/IEEE Std 242-1986, IEEE Recommended Practice for Protection and Coordination of Industrial and Commercial Power Systems.

[9] ANSI/NEMA ICS1-1983, General Standards for Industrial Control and Systems.

[10] ANSI/NEMA MG1-1978, Motors and Generators.

[11] ANSI/NFPA 30-1984, Flammable and Combustible Liquids Code.

[12] ANSI/NFPA 70-1987, National Electrical Code.

[13] ANSI/NFPA 110-1985, Emergency and Standby Power Systems.

[14] ANSI/UL 1008-1983, Safety Standard for Automatic Transfer Switches.

[15] CSA C22.2/178-1978, Automatic Transfer Switches.[17]

[16] EGSA 101P-1985, Engine Driven Generator Sets.

---

[17] In the US, CSA Standards are available from the Sales Department, American National Standards Institute, 1430 Broadway, New York, NY 10018. In Canada they are available at the Canadian Standards Association (Standards Sales), 178 Rexdale Blvd, Rexdale, Ontario, Canada M9W 1R3.

6.12 Bibliography

[B1] ANSI C97.1-1972 (R 1978), American National Standard Low-Voltage Cartridge Fuses 600 Volts or Less.

[B2] ANSI/IEEE 241-1983, IEEE Recommended Practice for Electric Power Systems in Commercial Buildings.

[B3] ANSI/NEMA ICS2-1983, Industrial Control Devices, Controllers and Assemblies.

[B4] ANSI/NFPA 99-1984, Health Care Facilities Code.

[B5] ANSI/UL 198B-1982, Safety Standard for Class H Fuses.

[B6] ANSI/UL 198C-1981, Safety Standard for High-Interrupting Capacity Fuses, Current-Limiting Types.

[B7] ANSI/UL 198D-1982, Safety Standard for Class K Fuses.

[B8] ANSI/UL 198E-1982, Safety Standard for Class R Fuses.

[B9] ANSI/UL 198H-1982, Safety Standard for Class T Fuses.

[B10] ANSI/UL 489-1985, Molded-Case Circuit Breakers and Circuit Breaker Enclosures.

[B11] CASTENSCHIOLD, R., GOLDBERG, D. L., JOHNSTON, F. C., and O'Donnell, P. *New Rating Requirements for Automatic Transfer Switches*. IEEE Conference Record, Paper no 78CH1302-91A, June 1978.

[B12] CASTENSCHIOLD, R. Plant Engineer's Guide to Understanding Automatic Transfer Switches. *Plant Engineering*, Mar 3, 1983.

[B13] *Surge Protection in Power Systems*. IEEE Tutorial Course Text, IEEE Paper no 79EH0144-6-PWR.

[B14] JOHNSON, G. S. Overcurrent Protection for Emergency and Standby Generating Systems. *Electrical Construction and Maintenance*, Apr 1983. See also: Stromme, G., Discussion, Aug 1983; Johnson, G. S., Discussion, Aug 1983; Phelps, E. J., Jr., Discussion, Nov 1983; Editor's Note, Nov 1983; Griffith, M. S., Discussion, Nov 1983; Johnson, G. S., Discussion, Feb 1984.

[B15] REICHENSTEIN, H. W. and CASTENSCHIOLD, R. Coordinating Overcurrent Protective Devices with Automatic Transfer Switches in Commercial and Institutional Power Systems. *IEEE Transactions on Industry Applications*, vol IA-11, no 6, Nov/Dec 1975.

Chapter 7
Grounding

7.1 Introduction

7.1.1 General. An ultimate goal of system protection is to prevent ground faults from occurring by proper design, installation, operation, and maintenance of electrical equipment and systems. However, some probability of ground faults due to accidents or insulation failure will exist in any electrical installation. Therefore, protection should be provided to safely clear the line-to-ground faults that may occur.

Reliable ground-fault protection for low-voltage electric systems requires coordinated design, proper installation, and routine maintenance of the following systems:

(1) Circuit protective equipment
(2) System grounding
(3) Equipment grounding

All references to the NEC in this chapter refer to ANSI/NFPA 70-1987 [4][18] (National Electrical Code).

7.1.2 Circuit Protective Equipment. Different types of circuit protection are satisfactory for ground-fault protection. Some commonly used types are fuses, circuit breakers with series trips, and circuit interrupting equipment tripped by ground-fault current sensing devices. Selection and application of circuit protective equipment requires a detailed analysis of each system and circuit to be protected, including the system and equipment grounding arrangements.

7.1.3 System and Equipment Grounding. This chapter presents basic system and equipment grounding requirements and grounding arrangements for emergency and standby power systems rated 600 V or less.

Factors that influence the selection of the type of system grounding, and the fundamental design principles for system and equipment grounding, are covered in detail in ANSI/IEEE Std 141-1986 [1] (IEEE Red Book) and ANSI/IEEE Std

---

[18] The numbers in brackets correspond to those of the references listed at the end of this chapter; when preceded by B, they correspond to the bibliography at the end of this chapter.

142-1982 [2] (IEEE Green Book). Ground-fault protection is also an important consideration which is covered in further detail in ANSI/IEEE Std 242-1986 [3] (IEEE Buff Book). The purpose of this chapter is to further define fundamental grounding requirements and to illustrate grounding arrangements for several types of system grounding and transfer schemes that are most often selected for emergency and standby power systems in industrial and commercial installations.

7.2 System and Equipment Grounding Functions

7.2.1 General. System grounding pertains to the manner in which a circuit conductor of a system is intentionally connected to earth, or to some conducting body that is effectively connected to earth or serves in place of earth. The following types of system grounding are discussed in this chapter:

(1) *Solidly Grounded.* A grounding electrode conductor connects a terminal of the system to the grounding electrode(s) and no impedance is intentionally inserted into the connection.

(2) *Resistance Grounded.* A grounding resistor is inserted into the connection between a terminal of the system and the grounding electrode(s).

(3) *Ungrounded.* None of the circuit conductors of the system are intentionally grounded.

Equipment grounding is the bonding together of all conductive enclosures for conductors and equipment in each circuit with equipment grounding conductors. These equipment grounding conductors are required to run with or enclose the circuit conductors, and they provide a permanent, low-impedance conductive path for ground-fault current. In solidly grounded systems, the equipment grounding conductors are bonded to the grounded circuit conductor and to the system grounding conductor(s) at specific points, as shown in Figs 66 and 67.

7.2.2 System Grounding Functions. Systems and circuit conductors are grounded to limit voltages due to lightning, line surges, or unintentional contact with higher voltage lines, and to stabilize the voltage to ground during normal operation. Systems and circuit conductors are solidly grounded to facilitate overcurrent device operation in case of ground faults.

Grounding electrodes and the grounding electrode conductors that connect the electrodes to the system grounded conductor are not intended to conduct ground-fault currents that are due to ground faults in equipment, raceways, or other conductor enclosures. In solidly grounded systems, the ground-fault current flows through the equipment grounding conductors from a ground fault anywhere in the system to the bonding jumper between the equipment grounding conductors and the system grounded conductor, as shown in Figs 67 and 68.

In solidly grounded service-supplied systems, the ground-fault current return path is completed through the bonding jumper in the service equipment and the grounded service conductor to the supply transformer, as shown in Fig 67.

7.2.3 Equipment Grounding Functions. Equipment grounding systems, consisting of interconnected networks of equipment grounding conductors, perform the following basic functions:

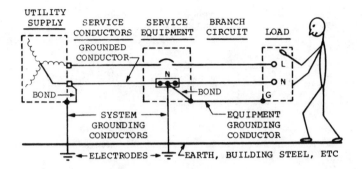

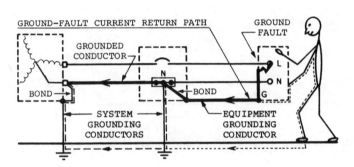

Fig 66
System and Equipment Grounding for
Solidly Grounded, Service-Supplied System

(1) They limit the voltage to ground (shock voltage) on the exposed noncurrent-carrying metal parts of equipment, raceways, and other conductor enclosures in case of ground faults

(2) They safely conduct ground-fault currents of sufficient magnitude for fast operation of the circuit protective devices

In order to ensure the performance of the above basic functions, equipment grounding conductors should:

(1) Be permanent and continuous

(2) Have ample capacity to conduct safely any ground-fault current likely to be imposed on them

(3) Have impedance sufficiently low to limit the voltage to ground to a safe magnitude and to facilitate the operation of the circuit protective devices

As illustrated in Figs 66 and 67, a person contacting a conductive enclosure in which there is a ground fault will be protected from shock injury if the equipment grounding conductors provide a shunt path of sufficiently low impedance to limit the current through the person's body to a safe magnitude. In solidly grounded systems, the ground-fault current actuates the circuit protective devices to automatically deenergize a faulted circuit and remove the shock exposure.

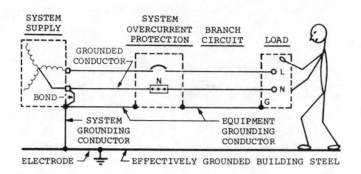

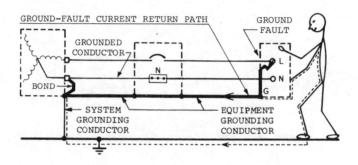

**Fig 67
System and Equipment Grounding for
Solidly Grounded, Separately Derived System**

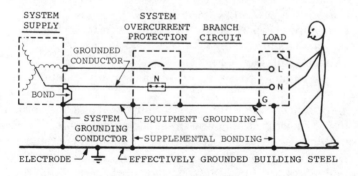

**Fig 68
Supplemental Equipment Bonding for Separately Derived System**

The same network of equipment grounding conductors should be provided for solidly grounded systems, high-resistance grounded systems, and ungrounded systems. Equipment grounding conductors are required in high-resistance grounded and ungrounded systems to provide shock protection and to present a low-impedance path for phase-to-phase fault currents in case the first ground fault is not located and cleared before another ground fault occurs on a different phase in the system.

7.3 Supplemental Equipment Bonding. Exposure to electrical shock can be reduced by additional supplemental equipment bonding between the conductive enclosures for conductors and equipment and adjacent conductive materials.

The supplemental equipment bonding shown in Fig 68 contributes to equalizing the potential between exposed noncurrent-carrying metal parts of the electric system and adjacent grounded building steel when ground faults occur. The inductive reactance of the ground-fault circuit will normally prevent a significant amount of ground-fault current from flowing through the supplemental bonding connections.

Ground-fault current will flow through the path that provides the lowest ground-fault circuit impedance. The ground-fault current path that minimizes the inductive reactance of the ground-fault circuit is through the equipment grounding conductors that are required to run with or enclose the circuit conductors. Therefore, practically all of the ground-fault current will flow through the equipment grounding conductors, and the ground-fault current through the supplemental bonding connections will be no more than required to equalize the potential at the bonding locations.

7.4 Objectionable Current Through Grounding Conductors. System and equipment grounding conductors should be installed and connected in a manner that will prevent an objectionable flow of current through the grounding conductors or grounding paths. If a grounded (neutral) circuit conductor is connected to the equipment grounding conductors at more than one point, or if it is grounded at more than one point, stray neutral current paths will be established. Stray neutral currents that flow through paths other than the intended grounded (neutral) circuit conductors during normal operation of a system will be objectionable if they contribute to any of the following:

(1) Interference with the proper operation of equipment, devices, or systems that are sensitive to electromagnetic interference, such as electronic equipment, communications systems, computer systems, etc

(2) Interference with the proper sensing and operation of ground-fault protection equipment

(3) Arcing of sufficient energy to ignite flammable materials

(4) Detonation of explosives during production, storage, or testing

(5) Overheating due to heat generated in raceways, etc, as a result of stray current

The grounded service conductor of service-supplied systems should be grounded at the service equipment. If the supply transformer is located outside of the building, the grounded service conductor should also be grounded on the secondary side of the supply transformer, either at the transformer or elsewhere ahead of the service equipment. A grounding connection should not be made to any grounded (neutral) circuit conductor on the load side of the service disconnecting means.

The stray neutral current illustrated in Fig 69 is due to multiple grounding of the grounded service conductor. Where there is a single overhead service drop connected to only one set of service entrance conductors, the stray neutral current shown in Fig 69 will normally be of insufficient magnitude to be objectionable.

Objectionable stray neutral currents are frequently caused by unintentional neutral-to-ground faults as shown in Fig 70. Neutral-to-ground faults are difficult to locate, but should be suspected if there are objectionable stray neutral currents.

Fig 69
Stray Neutral Current Due to Multiple
Grounding of Grounded Service Conductor

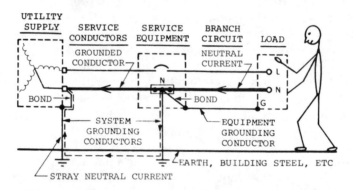

Fig 70
Stray Neutral Current Due to Unintentional
Grounding of a Grounded Circuit Conductor

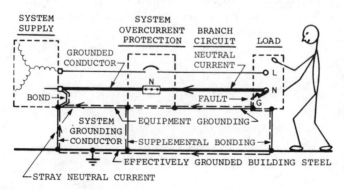

7.5 System Grounding Requirements. AC systems of 50–1000 V should be solidly grounded where the maximum voltage to ground on the ungrounded conductors will not exceed 150 V. Systems that supply phase-to-neutral loads should also be solidly grounded. The following commonly used systems are required to be solidly grounded:

(1) 120/240 V single-phase, three-wire
(2) 208Y/120 V three-phase, four-wire
(3) 480Y/277 V three-phase, four-wire
(4) 240/120 V three-phase, four-wire delta (midpoint of one phase used as a grounded circuit conductor)

The following commonly used systems are not required to be solidly grounded:

(1) 240 V delta, three-phase, three-wire
(2) 480 V three-phase, three-wire
(3) 600 V three-phase, three-wire

Where phase-to-neutral loads need not be served, there is a growing trend toward high-resistance grounding for 480 and 600 V three-phase systems in industrial establishments. High-resistance system grounding combines some of the advantages of solidly grounded systems and ungrounded systems. System overvoltages are held to acceptable levels during ground faults, and the potentially destructive effects of high-magnitude ground-fault currents that occur in high capacity, solidly grounded systems are eliminated. Ground-fault current is limited by the system grounding resistor to a magnitude that permits continued operation of a system while a ground fault is located and cleared. However, if a ground fault is not located and cleared before another ground fault occurs on another phase in the system, the high magnitude phase-to-phase fault current will flow through the equipment grounding conductors and operate the circuit protective equipment. Factors that influence selecting the type of system grounding are covered in the standards referenced in 7.13.

7.6 Types of Equipment Grounding Conductors. Equipment grounding conductors represented in the diagrams in this chapter may be any of the types permitted in the NEC [4], Section 250-91 (b), if this section is adopted by the authority having jurisdiction. Where raceway, cable tray, cable armor, etc, are used as equipment grounding conductors, all joints and fittings must be made tight to provide an adequate conducting path for ground-fault currents.

Earth and the structural metal frame of a building may be used for supplemental equipment bonding, but they should not be used as the sole equipment grounding conductor for ac systems. Equipment grounding conductors for ac systems should run with or enclose the conductors of each circuit.

Where copper or aluminum wire is used as equipment grounding conductors for circuits having paralleled conductors in multiple metal raceways, an equipment grounding conductor should be run in each raceway. The size of each paralleled equipment grounding conductor is a function of the rating of the circuit overcurrent protection.

7.7 Grounding for Separately Derived and Service-Supplied Systems. Basic grounding connections for solidly grounded separately derived and service-supplied systems are illustrated in Fig 71 with reference to NEC [4] sections in which the grounding requirements are specified.

The grounding requirements for separately derived systems and service-supplied systems are similar, but there are three important differences:

(1) The system grounded conductor for a separately derived system should be grounded at only one point. The single system grounding point is specified as the source of the separately derived system and ahead of any system disconnecting means or overcurrent devices. Where the main system disconnecting means is adjacent to the generator or transformer supplying a separately derived system, the grounding connection to the system grounded conductor may be made at, or ahead of, the system disconnecting means.

The system grounded conductor for a service-supplied system should be grounded at the service equipment and elsewhere on the secondary of the transformer that supplies the service, if the supply transformer is not in the same building as the service equipment.

(2) The preferred grounding electrode for a separately derived system is the nearest effectively grounded structural metal member of the structure or the nearest effectively grounded water pipe. The grounding electrode system for a service-supplied system should be in accordance with established code requirements.

(3) In solidly grounded separately derived systems, the equipment grounding conductors should be bonded to the system grounded conductor and to the grounding electrode conductor at or ahead of the main system disconnecting means or overcurrent device. The equipment grounding conductor should always be connected to the enclosure of the supply transformer or generator, as shown in Fig 71.

In solidly grounded service-supplied systems, the equipment grounding conductors should be bonded to the system grounded conductor and to the grounding electrode conductor at the service equipment. The grounded service conductor may be used to ground the noncurrent-carrying metal parts of equipment on the supply side of the service disconnecting means, and the grounded service conductor may also serve as the ground-fault current return path from the service equipment to the transformer that supplies the service.

7.8 Grounding Arrangements for Emergency and Standby Power Systems. A primary consideration in designing emergency and standby power systems is to satisfy the user's needs for continuity of electrical service. The type of system grounding that is employed, and the arrangement of system and equipment grounding conductors, will affect the service continuity. Grounding conductors and connections must be arranged so that objectionable stray neutral currents will not exist and ground-fault currents will flow in low-impedance, predictable paths that will protect personnel from electrical shock and assure proper operation of the circuit protective equipment.

Where phase-to-neutral loads must be served, systems are required to be solidly grounded. However, 600 and 480 V systems may be high-resistance grounded or

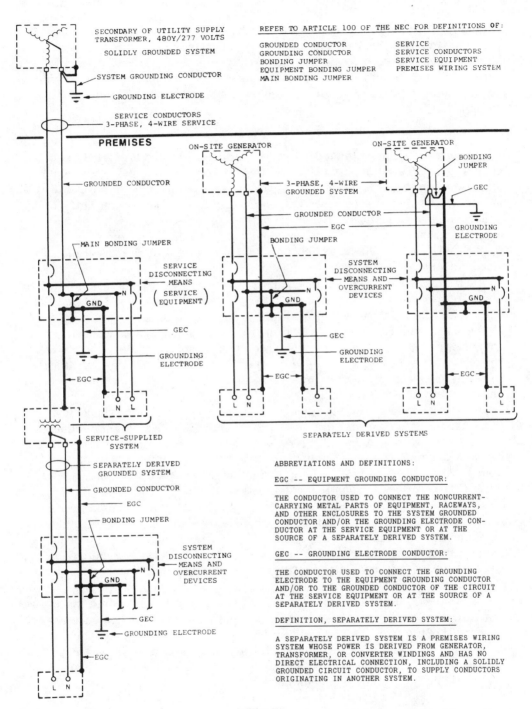

**Fig 71
System and Equipment Grounding for
Separately Derived and Service-Supplied Systems**

ungrounded where a grounded circuit conductor is not used to supply phase-to-neutral loads. High-resistance grounded or ungrounded systems may provide a higher degree of service continuity than solidly grounded systems.

This chapter discusses grounding arrangements for emergency and standby power systems that are solidly grounded, high-resistance grounded, and ungrounded.

7.9 Systems with a Grounded Circuit Conductor. Where grounded (neutral) conductors are used as circuit conductors in systems that have emergency or standby power supplies, the grounding arrangement should be carefully planned to avoid objectionable stray currents. For example, stray neutral currents and ground-fault currents in unplanned, undefined conducting paths may cause serious sensing errors by ground-fault protection equipment. A precautionary note is included in the NEC [4], Section 230-95, which states, "Where ground-fault protection is provided for the service disconnecting means and interconnection is made with another supply system by a transfer device, means or devices may be needed to assure proper ground-fault sensing by the ground-fault protection equipment."

7.9.1 Solidly Interconnected Multiple-Grounded Neutral. A grounded (neutral) circuit conductor is permitted to be solidly connected (not switched) in the transfer equipment. Therefore, a neutral conductor is permitted to be solidly interconnected between a service-supplied normal source and an on-site generator that serves as an emergency or standby source, as shown in Fig 72. However, this is not always a recommended practice.

Grounding connections to the grounded (neutral) conductor on the load side of the service disconnecting means is not recommended, so the grounding connection to the generator neutral in Fig 72 *should not be made.* Such multiple grounding of the neutral circuit conductor may cause stray currents that are likely to be objectionable and will cause ground-fault current to flow in paths that may adversely affect the operation of ground-fault protection equipment.

(1) *Figure 72.* The grounding connection to the neutral conductor at the on-site generator in Fig 72 is on the load side of the service disconnecting means, and is not recommended and may not satisfy code requirements. Where the grounded (neutral) conductor is solidly connected (not switched) in the transfer equipment, the system supplied by the on-site generator is not considered as a separately derived system.

(2) *Figure 73.* The grounding connection to the neutral conductor at the on-site generator in Fig 73 completes a conducting path for stray neutral current. The magnitude of the stray current will be a function of the relative impedances of the neutral current paths. Where ground-fault protection is provided at the service disconnecting means, the stray neutral current may adversely affect the operation of the ground-fault protection equipment.

(3) *Figure 74.* The grounding connection to the neutral, which is on the opposite side of the ground-fault sensor from the service entrance grounding, can cause ground-fault current to return through the neutral, thus cancelling the sensing of this current by the ground-fault sensor. It may also allow zero-sequence current to

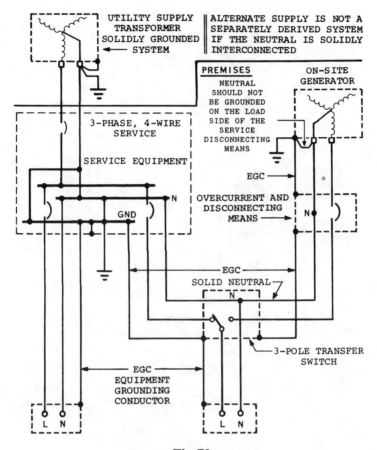

UTILITY SUPPLY
TRANSFORMER
SOLIDLY GROUNDED
SYSTEM

ALTERNATE SUPPLY IS NOT A
SEPARATELY DERIVED SYSTEM
IF THE NEUTRAL IS SOLIDLY
INTERCONNECTED

PREMISES
NEUTRAL
SHOULD NOT
BE GROUNDED
ON THE LOAD
SIDE OF THE
SERVICE
DISCONNECTING
MEANS

ON-SITE
GENERATOR

3-PHASE, 4-WIRE
SERVICE

SERVICE EQUIPMENT

EGC

N

GND

OVERCURRENT AND
DISCONNECTING
MEANS

N

EGC

SOLID NEUTRAL

N

3-POLE TRANSFER
SWITCH

EGC
EQUIPMENT
GROUNDING
CONDUCTOR

L N

L N

**Fig 72
Solidly Interconnected Neutral Conductor Grounded at
Service Equipment and at Source of Alternate Power Supply**

circulate in the neutral, even though no phase conductor is faulted, thus causing unwarranted tripping of the service entrance breaker.

If ground-fault protection is applied on the feeder to the transfer switch, it may be similarly affected. If neutral current continues to flow subsequent to tripping of the feeder breaker, it may damage the ground-fault relay. This flow of neutral current is not caused by the multiple grounds, they merely enable the flow. The current must be driven by some source, such as the induction due to unequal bus spacing, or even voltages from a different system.

Where ground-fault protection is applied to a feeder from the service equipment to the transfer equipment, the ground-fault current illustrated in Fig 74 could not be accurately detected by a zero-sequence ground-fault current sensor for the feeder.

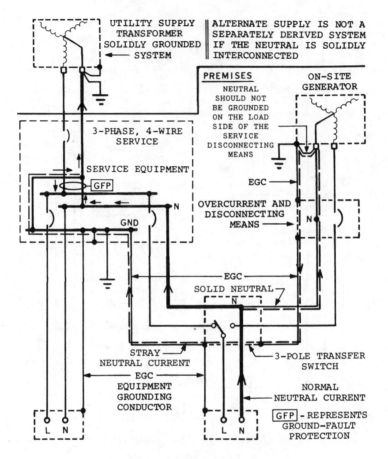

Fig 73
Stray Neutral Current Due to Grounding
the Neutral Conductor at Two Locations

(4) *Figure 75.* The multiple-grounding connections to the solidly interconnected neutral in Fig 75 permit a portion of the ground-fault current returning to the on-site generator to pass through the sensor for the ground-fault protection at the service equipment. This arrangement could result in tripping the service disconnecting means by ground-fault current supplies from the on-site generator.

7.9.2 Neutral Conductor Transferred by Transfer Means. The transfer switch may have an additional pole for switching the neutral conductor, or the neutral may be transferred by make-before-break overlapping neutral contacts in the transfer switch. Where the neutral circuit conductor is transferred by the transfer equipment, an emergency or standby system supplied by an on-site generator is a separately derived system. A separately derived system with a neutral circuit conductor should be solidly grounded at or ahead of the system disconnecting means.

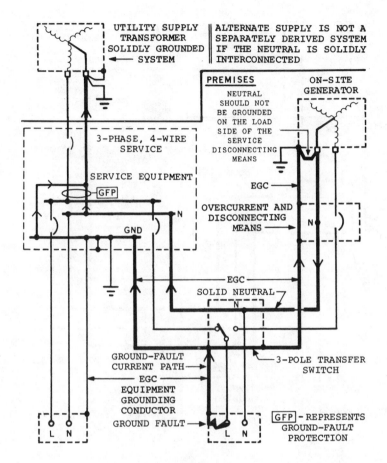

Fig 74
Ground-Fault Current Return Paths to Normal Supply,
Neutral Grounded at Two Locations

(1) *Figure 76.* The neutral conductor between the service equipment and the on-site generator in Fig 76 is completely isolated by the transfer switch and is solidly grounded at the service equipment and at the on-site generator. Where the neutral circuit conductor is switched by the transfer equipment, the stray neutral current paths and the undesirable ground-fault current paths illustrated in Figs 73, 74, and 75 are eliminated.

The normal supply and the alternate supply in Fig 76 are equivalent to two separate radial systems because all of the circuit conductors from both supplies are switched by the transfer switch. Since both systems are completely isolated from each other and are solidly grounded, ground-fault sensing and protection can be applied to the circuits of the normal source and the emergency or standby source as it is applied in a single radial system.

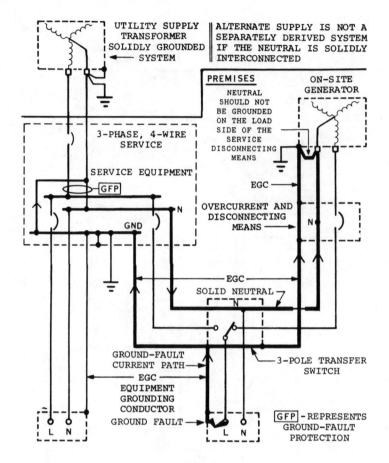

Fig 75
Ground-Fault Current Return Paths to Alternate Supply,
Neutral Grounded at Two Locations

(2) *Figure 77.* This diagram shows that unintentional neutral grounds will cause stray neutral currents that may be objectionable. Therefore, in addition to carefully planning the intentional grounding connections, systems should be kept free of unintentional grounds on the neutral circuit conductors.

(3) *Figure 78.* This diagram is similar to Fig 76 except there are two on-site generators connected for parallel operation and the neutral conductor is grounded at the generator switchgear instead of at the generator. Where the generator switchgear is adjacent to the generators, the grounding connections may be made in the generator switchgear, as shown in Fig 78.

Switching of the neutral conductor permits grounding of the neutral at the generator location. This may be desirable for the following reasons:

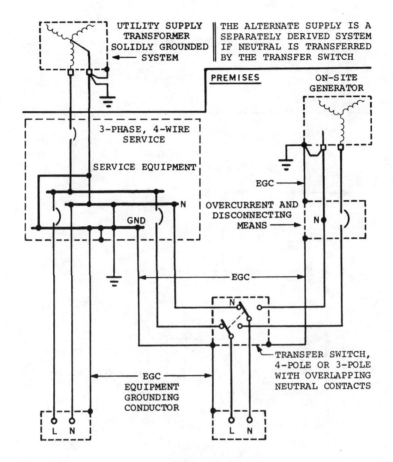

**Fig 76
Transferred Neutral Conductor Grounded at
Service Equipment and at Source of Alternate Power Supply**

(1) An engine-generator set is often remotely located from the grounded utility service entrance, and in some cases the ground potentials of the two locations may not be exactly the same.

(2) Good engineering practice requires the automatic transfer switch to be located as close to the load as possible to provide maximum protection against cable or equipment failures within the facility. The distance of cable between incoming service and the transfer switch and then to the engine-generator set may be substantial. Should cable failure occur with the neutral conductor not grounded at the generator location, the load would be transferred to an ungrounded emergency power system. Concurrent failure of equipment (breakdown between line and equipment ground) after transfer to emergency may not be detected.

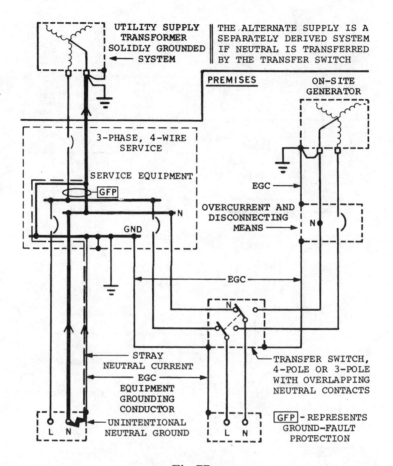

Fig 77
Stray Neutral Current Enabled by
Unintentionally Grounded Neutral Conductor

(3) Some local codes require ground-fault protection while the engine-generator is operating. This may present a sensing problem if the neutral conductor of the generator is not connected to a grounding electrode at the generator site and proper isolation of neutrals is not provided. This may also present a problem in providing a ground-fault signal as required per NEC [4], Section 700-7(e).

(4) When the transfer switch is in the emergency position, other problems may occur if the neutral is not grounded at the engine-generator set. A ground-fault condition, as shown in Fig 82, could cause nuisance tripping of the normal source circuit breaker, even though load current is not flowing through the breaker, and still not be cleared.

The normal neutral conductor and the emergency neutral conductor would then be simultaneously vulnerable to the same ground-fault current. The ground-fault

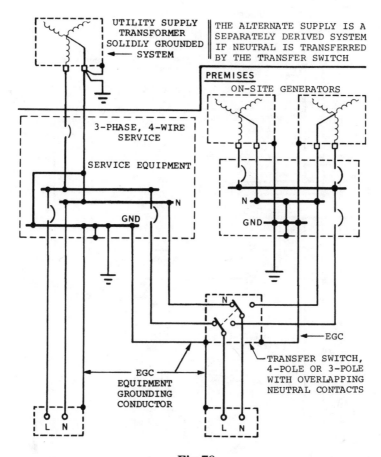

Fig 78
Transferred Neutral Conductor Grounded at Service Equipment
and at Switchgear for Two On-Site Generators Connected in Parallel

relay can be damaged. Thus a single fault could jeopardize power to critical loads even though both utility and emergency power are available. Such a condition may be in violation of codes requiring independent wiring and separate emergency feeders.

7.9.3 Neutral Conductor Isolated by a Transformer. Where a transferable load is supplied by a system that is derived from an on-site isolating transformer and the transfer equipment is ahead of the transformer, as illustrated in Fig 79, a grounded (neutral) circuit conductor is not required from either the normal or alternate supply to the transformer primary. The isolating transformer permits phase-to-neutral transferable loads to be supplied without a grounded (neutral) circuit conductor in the feeders to the transfer switch.

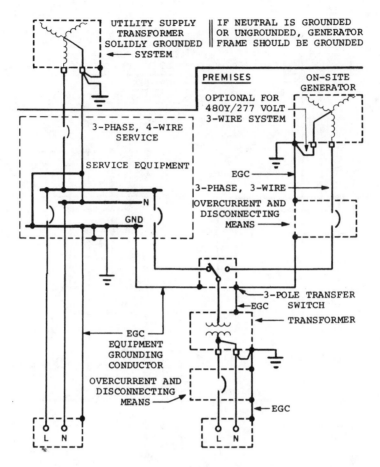

Fig 79
Solidly Grounded Neutral Conductor for
Transferred Load Isolated by Transformer

(1) *Figure 79.* The system supplied by the isolating transformer in Fig 79 is a separately derived system, and if it is required to be solidly grounded, it should be grounded in accordance with code requirements. The neutral circuit conductor for the transferable load in Fig 79 is supplied from the secondary of the isolating transformer.

If the on-site generator in Fig 79 is rated 480Y/277 V or 600Y/347 V, its neutral may not need to be solidly grounded because the neutral is not used as a circuit conductor. Therefore, the type of system grounding for such a generator is optional. However, the generator frame should be solidly grounded whether its neutral is ungrounded, high-resistance grounded, or solidly grounded.

(2) *Figure 80.* This figure illustrates ground-fault current return paths where the grounded (neutral) circuit conductor of a transferable load is isolated by a

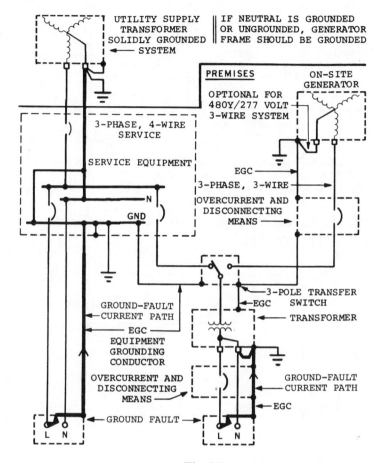

Fig 80
Ground-Fault Current Return Paths
Transferred Load Isolated by Transformer

transformer. Any stray neutral current or ground-fault current on the secondary of the isolating transformer will have no effect on ground-fault protection equipment at the service equipment or at the generator.

7.9.4 Solidly Interconnected Neutral Conductor Grounded at Service Equipment Only. Where the grounded (neutral) circuit conductor is solidly connected (not switched) in the transfer equipment, an emergency or standby system supplied by an on-site generator should not be considered a separately derived system. The solidly interconnected grounded (neutral) conductor needs only to be grounded at the service equipment.

(1) *Figure 81.* The solidly interconnected neutral conductor in Fig 81 is grounded at the service equipment only and there are no conducting paths for stray neutral currents. If the generator is not adjacent to the service equipment, the

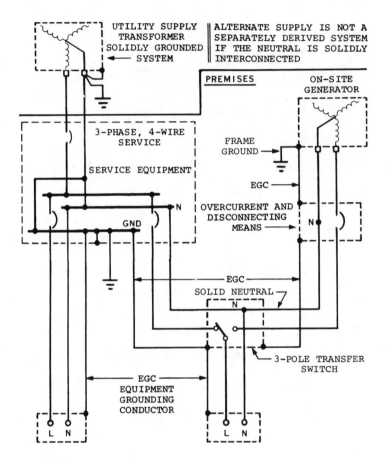

Fig 81
Solidly Interconnected Neutral Conductor
Grounded at Service Equipment Only

generator frame should be connected to a grounding electrode such as effectively grounded building steel.

(2) *Figure 82*. Where the solidly interconnected neutral conductor is grounded at the service equipment only, as shown in Fig 82, the ground-fault current return path from the transferable load to the on-site generator is through the equipment grounding conductor from the transfer switch to the service equipment, then through the main bonding jumper in the service equipment and the neutral conductor from the service equipment to the generator.

The ground-fault current shown in Fig 82 might trip the service disconnecting means, even though the ground fault is on a circuit supplied by the generator. A signal could be derived from a ground-fault sensor on the generator neutral conductor to block the ground-fault protection equipment at the service in case of

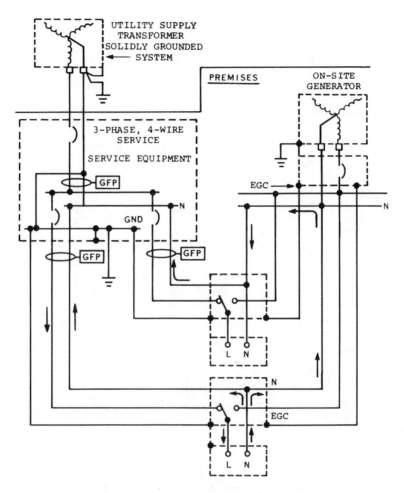

Fig 82(a)
Two Transfer Switches with Multiple Neutral Return
Paths — Ground Fault Indicated Though None Exists

ground faults while the system is transferred to the generator. Such blocking signals require careful analysis to ensure proper functioning of the ground-fault protection equipment.

If the neutral conductor between the service equipment and the transfer switch in Fig 82 is intentionally or accidentally disconnected, the generator will be ungrounded. Therefore, the integrity of the neutral conductor should be maintained from the service equipment to the transfer switch while the load is transferred to the generator. The equipment grounding conductor should also be maintained from the service equipment to the transfer equipment in order to provide a ground-fault current return path from the transferable load to the generator.

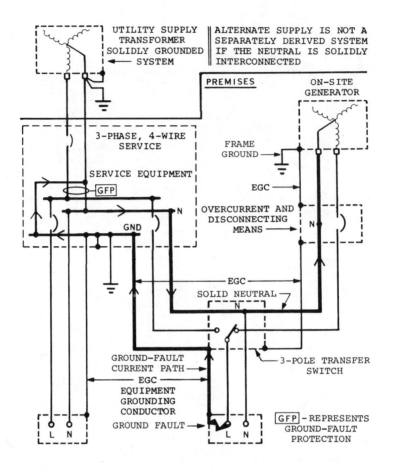

Fig 82(b)
Ground-Fault Current Return Path to Alternate Supply,
Neutral Conductor Grounded at Service Equipment Only

7.9.5 Multiple Transfer Switches. For increased reliability, multiple transfer switches, located close to the loads, are often used rather than one transfer switch for the entire load. In such cases, consideration should be given to the possibility of cable or equipment failure between the service equipment and the transfer switches, thus possibly causing an emergency or standby power system to become ungrounded. This is particularly important if a solidly interconnected neutral conductor is grounded at the service equipment only.

When multiple transfer switches are used with ground-fault detection on the distribution circuit level, there is a possibility of tripping the ground-fault circuits when no ground fault exists. Figure 82(a) illustrates a typical system with multiple 3-pole transfer switches. Power is supplied from the normal source to the load through a transfer switch; the neutral current returns to the transfer switch where multiple return paths are presented. The current will divide between the paths depending on the relative impedance of the path. Some portion of the current will return by the normal return path. The remainder of the current will flow through the neutral conductor of the emergency supply to the transfer switch, through the neutral conductor of the emergency distribution bus, through the neutral conductor of the emergency supply to the second transfer switch, then through the neutral conductor of the normal supply to the second transfer switch to the neutral conductor of the normal distribution bus, and finally back to the normal source.

This situation can occur independently of the grounding means chosen for the system, and can cause tripping of both normal and emergency service distribution breakers if they are protected by ground-fault devices even when no ground-fault exists. This problem can be corrected by the use of the 3-pole transfer switches with overlapping neutral or 4-pole transfer switches.

7.9.6 Multiple Engine-Generator Sets. When several engine-generator sets are connected in parallel and serve as a common source of power, each generator neutral is usually connected to a common neutral bus within the paralleling switchgear which, in turn, is grounded. The associated switchgear containing the neutral bus should be located in the vicinity of the generator sets. A single system grounding conductor between the neutral bus and ground simplifies the addition of ground-fault sensing equipment. This also permits the use of a single grounding resistor for multiple engine-generator sets.

It may be argued when individual grounding resistors are used, circulation of harmonic currents between paralleled generators is not a problem, since the resistance limits the circulating current to negligible values. However, if third harmonics are suppressed in the engine-generator sets, circulating currents are usually not a problem.

Multiple engine-generator sets that are physically separated and used for isolated loads may necessitate additional neutral-to-ground connections. However, by using multiple 4-pole transfer switches or 3-pole switches with overlapping neutral contacts, proper isolation and ground-fault sensing can be obtained.

7.9.7 Transferring Neutral Conductor. The foregoing discussion and circuit diagrams indicate 4-pole transfer switches are synonymous with 3-pole transfer switches with overlapping neutral contacts. This is true from the standpoint of providing isolation as needed for ground-fault protection. However, there are other considerations.

Four-pole transfer switches can be satisfactorily applied where the loads are passive and relatively balanced. However, unbalanced loads may cause abnormal voltages for as long as 10–15 ms when the neutral conductor is momentarily opened during transfer of the load. Transfer switches may be called upon to operate during total load unbalance caused by a single-phasing condition. Inductive loads may cause additional high transient voltages in the microsecond range. Contacts of the fourth switch pole do interrupt current and are, therefore, subject to arcing and contact erosion. A good maintenance program is recommended to reaffirm at intervals the integrity of the fourth pole as a current-carrying member with sufficiently low impedance.

With overlapping neutral contacts, the only time the neutrals of the normal and emergency power sources are connected is during transfer and retransfer. The operating time of the ground-fault relay should be set higher than the overlap time of the particular transfer switch used.

With overlapping neutral transfer contacts, the load neutral is always connected to one source of power. When the transfer switch operates, abnormal and transient voltages are kept to a minimum because there is no momentary opening of the neutral conductor. Also, there is no erosion of the overlapping contacts due to arcing.

7.10 Ground-Fault Alarm. Most of the discussion on ground faults in this chapter has pertained to protection on the normal side of the transfer switch. However, ground faults can also occur between the transfer switch and the on-site generator. It is usually not desirable that emergency loads be automatically disconnected from the on-site generator should such a ground-fault occur.

However, for the purpose of safety and to minimize the possibility of fire and equipment damage, a visual and audible alarm to indicate a ground fault in larger emergency power systems is often recommended. This could alert personnel that corrective action should be taken. Instructions should be provided at or near the sensor location stating what course of action is to be taken in the event of a ground-fault indication.

Usually a ground-fault condition will show up during routine testing of the on-site generator. This then provides time to take corrective steps prior to an actual emergency condition.

7.11 Systems Without a Grounded Circuit Conductor. Three-phase, three-wire, 480 and 600 V systems, which are extensively used in industrial establishments, do not require the use of grounded conductors as circuit conductors. There are more system grounding options where emergency and standby power systems do not require a grounded circuit conductor to supply phase-to-neutral loads.

7.11.1 Solidly Grounded Service. In many installations the service to the premises will be solidly grounded, three-phase, four-wire where a grounded (neutral) circuit conductor is not required for loads that are provided with an on-site emergency or standby supply. An on-site emergency or standby supply is not always required to have the same type of system grounding as the normal supply to the premises.

(1) *Figure 83*. The generator in Fig 83 supplies a three-phase, three-wire system. If the generator is rated 480Y/277 V or 600Y/347 V, it need not be used as a circuit conductor.

An on-site generator that is not required to be solidly grounded may be high-resistance grounded or ungrounded. A high-resistance grounded or ungrounded emergency or standby power supply provides a high degree of service continuity

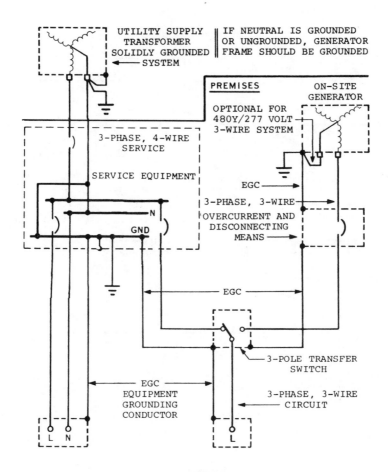

Fig 83
3-Pole Transfer Switch for Transfer to an Alternate
Power Supply Without a Grounded Circuit Conductor

because the circuit protective equipment will not be tripped by the first ground fault on the system.

If the generator in Fig 83 is solidly grounded, it should be grounded at or ahead of the generator disconnecting means.

(2) *Figure 84.* Interlocked circuit breakers are used as the transfer means in Fig 84. The grounding arrangements in Figs 83 and 84 are the same. Where a grounded (neutral) circuit conductor is not required, the type of transfer equipment that is employed is not a consideration in selecting the type of system grounding for the emergency or standby supply.

Interlocked circuit breakers should not be selected as a transfer means without considering additional means to isolate the normal and alternate circuit conductors and equipment for maintenance work. The overall reliability of a system may be reduced due to additional exposure where the circuit conductors from the

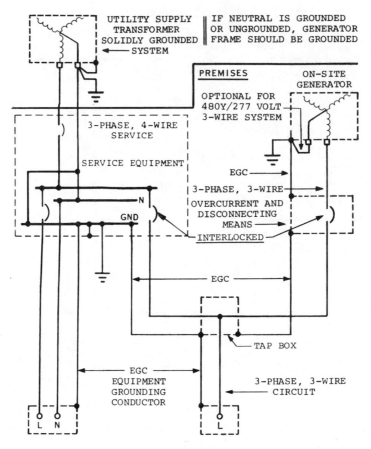

Fig 84
Interlocked Circuit Breakers for Transfer to an Alternate
Power Supply Without a Grounded Circuit Conductor

normal supply are solidly connected to the circuit conductors from the emergency or standby supply. The circuit wiring for an emergency system should be kept entirely independent of all other wiring or equipment, except in transfer switches or in junction boxes and in fixtures for exit and emergency lighting.

(3) *Figure 85*. A 480 or 600 V, three-phase, three-wire, on-site generator that is high-resistance grounded may serve as an emergency or standby power supply for a three-phase, three-wire system that is normally supplied by a solidly grounded service, as shown in Fig 85.

7.11.2 High-Resistance Grounded Service. Where the three-phase, three-wire critical load is relatively large compared with loads that require a grounded (neutral) circuit conductor, a high-resistance grounded service with a high-resistance grounded emergency or standby power supply is sometimes considered. This arrangement requires an on-site transformer for loads that require a neutral

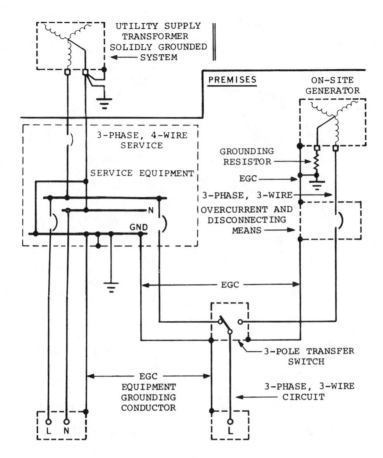

Fig 85
High-Resistance Grounded Alternate Power
Supply Without a Grounded Circuit Conductor

circuit conductor. If an on-site transformer is provided for emergency lighting, it should be supplied from the normal service and the emergency supply through automatic transfer equipment.

(1) *Figure 86.* The supply transformer for normal service and the on-site generator in Fig 86 are both high-resistance grounded. There are no provisions in this diagram to supply phase-to-neutral loads.

(2) *Figure 87.* The ground-fault current return path to the on-site generator in Fig 87 is completed through the grounding resistor. The grounding resistor limits the line-to-ground fault current to a magnitude that can be tolerated for a time, allowing the ground fault to be located and removed from the system.

High-resistance grounded systems should not be used unless they are equipped with ground-fault indicators or alarms, or both, and qualified persons are available to quickly locate and remove ground faults. If ground faults are not promptly

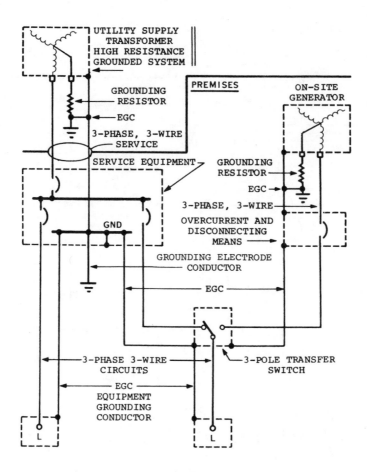

Fig 86
High-Resistance Grounded Systems
for Normal Service and Alternate Supply

removed, the service reliability will be reduced. A possible aid in locating ground faults is applying a pulsed current that can be traced to the fault.

7.12 Mobile Engine-Generator Sets. The basic requirements for system grounding and equipment grounding that are discussed and illustrated in this chapter also apply to mobile engine-generator sets when they are used to supply emergency or standby power for systems rated 600 V or less. Grounding arrangements that are acceptable for emergency or standby power supplied by fixed generators are also acceptable where such systems are supplied from mobile engine-generators. If ground-fault protection equipment is used for the normal service or for a mobile generator that supplies emergency or standby power, the system and equipment grounding connections should be arranged to prevent stray neutral currents and

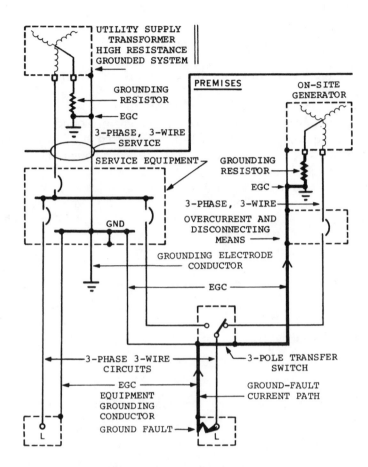

Fig 87
Ground-Fault Current Return Path to
High-Resistance Grounded Alternate Supply

ground-fault currents in paths that cause improper operation of the ground-fault protection devices.

Where mobile generators are used to supply emergency or standby power for systems similar to Fig 76 in which the neutral is grounded at the generator, it may be desirable to install permanent, pretested system grounding electrodes at locations designated for operating the mobile equipment. If mobile generators are operated at locations where permanent grounding electrodes are not available, and the generator neutral is required to be grounded, plate electrodes should be considered because they can be installed quickly. Plate electrodes for equipment grounding should be located close to the generator and should be placed on the surface (ground) with sufficient hold-down weight to ensure electrical bypass between the operator's feet and the generator frame. However, this approach should not detract from the desirability of connecting the generator neutral to a permanent grounding electrode.

Equipment grounding conductors should be provided with the mobile generator circuit conductors as illustrated in the figures in this chapter. If flexible cords or cables are used to connect mobile generators to emergency or standby power systems, each cord or cable should have an equipment grounding conductor.

Supplemental equipment bonding, which is discussed in 7.3, will reduce the risk of electrical shock for persons who contact the exposed noncurrent-carrying metal parts of the mobile equipment. Supplemental equipment bonding should be installed between the mobile generator frame and adjacent conductive surfaces such as structural steel, metal piping systems, and metal equipment enclosures. If the neutral of a mobile generator is not grounded, the generator frame should be connected to a grounding electrode, as shown in Fig 82, in addition to its connection to the equipment grounding conductor network.

7.13 References. The following publications shall be used in conjuction with this chapter.

[1] ANSI/IEEE Std 141-1986, IEEE Recommended Practice for Electric Power Distribution for Industrial Plants.

[2] ANSI/IEEE Std 142-1982, IEEE Recommended Practice for Grounding of Industrial and Commercial Power Systems.

[3] ANSI/IEEE Std 242-1986, IEEE Recommended Practice for Protection and Coordination of Industrial and Commercial Power Systems.

[4] ANSI/NFPA 70-1987, National Electrical Code.

[5] NEMA PB 2.2-1983, Application Guide for Ground Fault Protective Devices for Equipment.[19]

7.14 Bibliography

[B1] BLOOMQUIST, W. C., OWEN, K. J., and GOOCH, R. L. High-Resistance Grounded Power Systems—Why Not. *IEEE Transactions on Industry Applications*, vol IA-12, Nov/Dec 1976, pp 574–579.

[B2] BRINKS, J. W. and WOLFINGER, J. P., JR. Systems Grounding, Protection, and Detection for Cement Plants. *IEEE Transactions on Industry Applications*, vol IA-17, Nov/Dec 1981, pp 587–596.

[B3] CASTENSCHIOLD, R. Grounding of Alternate Power Sources. *Conference Record of the 1977 IEEE Industry Applications Society, Technical Conference*, paper 77CH11246-8-IA, pp 67–72.

---

[19] NEMA publications can be obtained from the National Electrical Manufacturers Association, 2101 L Street NW, Washington, DC 20037.

[B4] CASTENSCHIOLD, R. Ground-Fault Protection of Electrical Systems With Emergency or Standby Power. *IEEE Transactions on Industry Applications*, vol IA-13, Nov/Dec 1977, pp 517–523.

[B5] KAUFMANN, R. H. Some Fundamentals of Equipment-Grounding Circuit Design. *AIEE Transactions (Applications and Industry)*, Pt II, vol 73, Nov 1954, pp 227–232.

[B6] RICHARDS, J. H. *Generator Neutral Grounding, Electrical Construction, and Maintenance*, May 1984, p 122.

[B7] WEST, R. B. Equipment Grounding for Reliable Ground Fault Protection in Electrical Systems Below 600 V. *IEEE Transactions on Industry Applications*, vol IA-10, Mar/Apr 1974, pp 175–189.

[B8] WEST, R. B. Grounding for Emergency and Standby Power Systems. *IEEE Transactions on Industry Applications*, vol IA-15, Mar/Apr 1979, pp 124–136.

[B9] ZIPSE, D. W. Multiple Neutral to Ground Connections. *Conference Record of the 1972 IEEE Industrial and Commercial Power Systems, Technical Conference*, paper 72CH0600-7-IA, pp 60–64.

Chapter 8
Maintenance

8.1 Introduction. Once it has been determined through reliability analysis that an emergency or standby power system is justified, available systems should be evaluated in order to select one that will economically satisfy the application requirements. Selection of the proper system requires that the installation practices and maintenance procedures are given sufficient emphasis. After the system has been selected and responsibility established, the plans for preventive maintenance should be completed. Preventive maintenance for electrical equipment consists of a system of planned inspections, testing, cleaning, drying, monitoring, adjusting, corrective modification, and minor repair to maintain equipment in optimum operating condition and with maximum reliability. Without proper emphasis on maintenance from the design stage through to operation, the system can rapidly become unreliable and fail to meet the intended design goal. This chapter presents a consolidation of preventive maintenance recommendations.

The goal of preventive maintenance plans is to ensure that equipment is in optimum operating condition. Because the equipment is part of an emergency and standby system, it becomes more important and a greater challenge. Preventive maintenance now becomes a science of anticipation and prediction of failures. In this case, the following vital precautions should be taken:

(1) Make sure that the installation will not be subject to ventilation problems, or obstructed by junk or storage.

(2) Place regular test responsibilities on trained personnel and schedule tests frequently to assure operation when required.

(3) Gasoline and, to a lesser extent, diesel fuels deteriorate when stored for extended periods of time. Inhibitors can be used to reduce the rate of deterioration, but it is sound practice to operate a system utilizing these fuels such that total operating time will result in a complete fuel change cycle every few months.

Although many different emergency and standby systems are on the market today, the general maintenance recommendations presented will provide an indication of preventive maintenance needs for various systems. The recommendations are broken down into system components. The components can be reconstructed to form particular systems.

More recently, various codes have stressed the need for routine maintenance and operational testing of emergency and standby systems. They require that a written record of inspections and repairs be maintained on the premises. For further information see ANSI/NFPA 99-1984 [1][20] and ANSI/NFPA 110-1985 [2].

8.2 Internal Combustion Engines. The types of internal combustion engines commonly available include natural and bottled gas, gasoline, and diesel. Since there are more similarities than differences in their maintenance requirements, engines have been discussed as a group.

Long life and high reliability are characteristics to be expected from these types of prime movers, but only if properly maintained. Preventive maintenance programs will greatly contribute to service life and reliability.

In establishing a preventive maintenance program for these engines, the best starting point is the manufacturer's service manual. This will provide a guide for specific points to be checked and indicate the frequency of inspection. These reference points can be modified to fit particular installation and operating conditions.

More than any other factor, lubrication determines an engine's useful life. Various parts of the engine may require different lubrications and different frequencies of lubricant application. It is important to follow the manufacturer's recommendations as to type and frequency of lubrication.

Cleanliness should be the foundation of a preventive maintenance program. While there may be minimal wear, there is always a possibility of contamination by corrosive dirt and grit buildup. Dirt is a major cause of equipment failure. Before performing any inspection or service, carefully clean all fittings, caps, filler and level plugs, and their adjacent surfaces to prevent contamination of lubricants and coolants.

Routinely scheduled inspection areas should include radiator coolant level, antifreeze if utilized, crankcase oil level, fuel supply, and air cleaner. A drain check should be made to eliminate condensed water from fuel tank and filters. The engine should be visually inspected to detect loose nuts, bolts, and other hardware, and leaks at seals, gaskets, and other connections in the fuel, cooling, lubrication, and exhaust systems.

8.2.1 Typical Maintenance Schedule. The following maintenance schedule is not intended as a recommendation, but is presented as a typical service guide for an 1800 r/min unit.

Every 25 h of operation (or 4 months):
  (1) Adjust fan and alternator belt
  (2) Add oil to cup for distributor housing
  (3) Change oil in oil-type air filter

Every 50 h of operation (or 6 months):
  (1) Drain and refill crankcase
  (2) Clean crankcase ventilation air cleaner

---

[20] The numbers in brackets correspond to those of the references listed at the end of this chapter; when preceded by B, the correspond to the bibliography at the end of this chapter.

(3) Clean dry-type air cleaners
(4) Check transmission oil
(5) Check battery
(6) Clean external engine surface
(7) Perform 25 h service (above)

Every 100 h of operation (or 8 months):
(1) Replace oil filter element
(2) Check crankcase ventilator valve
(3) Clean crankcase inlet air cleaner
(4) Clean fuel filter
(5) Replace dry-type air cleaner
(6) Perform 25 and 50 h service (above)

Every 200 h of operation (or 12 months):
(1) Adjust distributor contact points
(2) Check spark plugs for fouling and proper gap
(3) Check timing
(4) Check carburetor adjustments
(5) Perform 25, 50, and 100 h service (above)

Every 500 h of operation (or 24 months):
(1) Drain and refill transmission
(2) Replace crankcase ventilator valve
(3) Replace one-piece-type fuel filter
(4) Check valve-tappet clearance
(5) Check crankcase vacuum
(6) Check compression
(7) Perform 25, 50, 100, and 200 h service (above)

8.3 Gas Turbine

8.3.1 General. The combustion gas turbine, as with any rotating power equipment, requires a program of scheduled inspection and maintenance to achieve optimum availability and reliability. The combustion gas turbine is a complete, self-contained prime mover. This combustion process requires operation at high temperatures. When inspection marking on stainless steel parts is necessary, a grease pencil should be used. Graphite particles from lead pencils will carburize stainless steel at the high temperature of gas turbine operation.

Starting reliability is of prime concern since a delay in starting usually means the need for the unit has passed.

8.3.2 Operating Factors Affecting Maintenance. The factors having the greatest influence on scheduling of preventive maintenance are type of fuel, starting frequency, environment, and reliability required.

8.3.2.1 Fuel. The effect of the type of fuel on parts is associated with the radiant energy in the combustion process and the ability to atomize the fuel. Natural gas, which does not require atomization, has the lowest level of radiant energy and will produce the longest life of parts. Diesel fuels will provide the next longest life, and crude oils and residual oils, with higher radiant energy and more difficult atomization, will provide shorter life of parts.

Contaminants in the fuel will also affect the maintenance interval. In liquid fuels, dirt results in accelerated wear of pumps, metering elements, and fuel nozzles. Contamination in gas fuel systems can erode or corrode control valves and fuel nozzles. Filters must be inspected and replaced to prevent carrying contaminants through the fuel system. Clean fuels will result in reduced maintenance and extended lives of parts.

8.3.2.2 Starting Frequency. Each stop and start subjects a gas turbine to thermal cycling. This thermal cycling will cause a shortened life of parts. Applications requiring frequent starts and stops dictate a shorter maintenance interval.

8.3.2.3 Environment. The condition of inlet air to a combustion gas turbine can have a significant effect on maintenance. Abrasives in the inlet air, such as ash particles, require that careful attention be paid to inlet filtering to minimize the effect of the abrasives. In the case or corrosive atmosphere, careful attention should be paid to inlet air arrangement and the application of correct materials and protective coatings.

8.3.2.4 Reliability Required. The degree of reliability required will affect the scheduling of maintenance. The higher the reliability desired, the more frequent the maintenance is required.

8.3.3 Typical Maintenance Schedule. The following maintenance schedule is included to show typical maintenance requirements for a combustion gas turbine in an emergency or standby power application.

Weekly:

(1) Check oil level

(2) Check for adequate fuel pressure

(3) Visually inspect all nuts and other fasteners

(4) Check for oil or fuel leakage

(5) Check electrical connections for tightness and corrosion

(6) Check all lines and hoses for external wear

(7) Check air inlet for obstructions

(8) Check exhaust for obstructions

(9) Bleed all drains and collectors to detect obstructions

Every 250 h of operation:

(1) Replace oil filter element

(2) Replace fuel filters as necessary

(3) Check batteries

(4) Blow out air lines and filters with dry low-pressure air

(5) Lubricate auxiliary motors

Every 1000 h of operation:

(1) Inspect spark plug

(2) Inspect fuel injectors and combustion parts

(3) Blow low-pressure dry air through exhaust and combustor drain lines

(4) Inspect entire engine for unusual discoloration, cracks, wear, or chafing of hoses, lines, and wires, and other unusual operating conditions

(5) Check condition of air inlet filters

(6) Check condition of engine and exhaust thermocouples

(7) Check temperature and speed control units

The manufacturer's service manual should be used as a starting point for establishing a preventive maintenance program for combustion gas turbines. Adjustments in maintenance intervals can be made from records or operating experience.

8.4 Generators. Keeping equipment clean is of primary importance in generator preventive maintenance. Dust, oil, moisture, or other substances should not be allowed to accumulate on the equipment. Ventilation ducts should be kept clean to allow maximum cooling air into the generator. The importance of keeping windings clean cannot be overemphasized. Dust, dirt, and other foreign matter can restrict heat dissipation and deteriorate insulation. A layer of dust as thin as 30 mils can raise the operating temperature of generator windings 10 °C.

The best method for cleaning a generator of loose and dry particles is to use a vacuum cleaner with proper fittings. Blowing with 30 lb/in^2 compressed air can also be employed, but this method has a tendency to redeposit the particles. Wiping with a soft, clean rag has the disadvantage of not being able to remove dust from grooves and inaccessible places. Buildup of grease and oil can be removed by conservative use of solvents. Megohmmeter readings should be taken after cleaning and drying. If the resistance is too low, the cleaning should be repeated.

Regularly scheduled inspections should include checking for tightness, checking of all wires for chafed, brittle, or otherwise damaged insulation; and checking bearings, brushes, and commutator for proper operating condition. Inspect for creepage of bearing grease inside generator. If moisture has accumulated in the generator, the unit should be dried and strip heaters or other methods should be utilized to prevent the condition from recurring. To dry the unit, external heat should be applied to reduce the moisture content. Internal heat may then be applied by introducing a low-voltage current through the windings. The winding temperature must be monitored to prevent insulation damage during the drying operation.

Brushes and connecting shunts should be inspected for wear and deterioration. Remove the brushes one at a time and check for length. Be sure the brushes move freely in their holders. Brush holders should be checked for proper tension. If the spring tension is strong and is not adjustable, the brush holder should be replaced.

Brushes should be replaced when worn down to ½ in. Replace brushes in complete sets, not a single brush. Be sure the shunt leads are properly connected. After placing new brushes in the holders, carefully fit the contact surface of the brushes to the commutator by using first #1, then #00 sandpaper. Do not use emery cloth. Cut the sandpaper into strips slightly wider than one brush. Insert the strip under the brush with the smooth side toward the commutator and draw back and forth around the commutator in the manner of a slipping belt. After the brushes have been seated, clean the carbon dust from the commutator and brush assemblies.

Commutators should be smooth and have a light to medium brown color. A rough or blackened generator commutator may be polished with a commutator dressing stone fitted to the curvature of the commutator. If not available, use #00 sandpaper with a block of wood shaped to fit the curvature of the commutator.

Do not use emery cloth. All brushes should be lifted and the generator driven while slowly moving the polishing block back and forth. The mica insulation between commutator bars should be undercut $1/16$–$1/32$ in. As the commutator wears down, the mica will cause ridges resulting in bouncing of the brushes. If this condition exists, the mica must be undercut and the commutator resurfaced by a qualified repairman. Do not use lubricants of any kind on the commutator.

Generator bearings should be subjected to careful inspection at regularly scheduled intervals. The frequency of inspection, including addition or changing of oil or grease, is best determined by a study of the particular operating conditions. Sealed bearings require no maintenance and should be replaced when worn or loose. When the generator is running, listen for unusual noise and feel the bearing housing for vibration or excessive heat.

With proper preventive maintenance a generator will provide reliable and lengthy service.

8.5 Static Uninterruptible Power Supplies. Most uninterruptible power supplies (UPSs) are installed because the load requires an uninterruptible source of power. Therefore, special precautions must be taken when isolating a UPS for maintenance. Be aware of the loads supplied by the UPS and notify necessary personnel that disconnection of the UPS is required for maintenance.

A static UPS is extremely reliable and requires very little maintenance of the inverter and battery charger. See 5.6 for maintenance of batteries.

The UPS should be completely isolated from input and output power, including batteries. Many UPSs may have a unique isolating procedure to prevent an outage to a critical load.

Preventive maintenance consists of cleaning and inspecting the UPS at regular periodic intervals. In dirty environments, the time between intervals should be decreased. The following items should be included in the regular scheduled inspection and service schedule:

(1) Remove all input and output power from the unit before performing preventive maintenance work

(2) Discharge and ground all capacitor terminals in the charger and inverter with a grounding stick

(3) Inspect all parts for evidence of overheating

(4) Inspect all parts for evidence of physical damage

(5) Inspect terminals for loose or broken connections, or burned insulation

(6) Inspect and clean, as necessary, all air intakes and exhaust openings

(7) Check for liquid contamination (battery electrolyte, oil from capacitors, etc)

(8) Tighten all terminations

(9) Check battery condition

(10) Connect source power and check control circuit power supply voltages per manufacturer's specifications

(11) Check and adjust voltage output and the frequency per manufacturer's specifications

After the UPS has been reconnected, check the output voltage and frequency under load. Simulate a power failure and check for proper system operation.

8.6 Batteries

8.6.1 General. All emergency and standby power systems are dependent upon proper battery and associated charger operation. Cleanliness, electrolyte level, voltage, and specific gravity records combined with conscientious visual inspections are essential to proper battery and charger maintenance.

During use and normal charging, the loss of electrolyte will be minimum and water alone should be added to correct the electrolyte level drop due to evaporation. Water addition is the single most important maintenance requirement. In some locations, tap water may be used, but in most localities, it is best to use distilled or demineralized water. Care should be taken to avoid bringing the solution level to any higher than the top mark on the cell. Overflow can occur during gassing if too much water was added. Under no circumstances should the electrolyte be allowed to drop below the top of the separator in the cell's elements. A reduction of the cell's discharge capacity will result, and permanent damage to the plates may result as well.

8.6.2 Safety. Hydrogen is liberated from all wet electrolyte-type batteries during the charging period. Therefore, adequate ventilation by dilution or mechanical means should be provided to avoid an explosive mixture. A "no smoking" sign should be posted in immediate battery areas. In addition, protective equipment should be worn to prevent contact with battery acid.

8.6.3 Battery Types. The major types of wet electrolyte batteries are lead–antimony, lead–calcium, and nickel–cadmium. There is also the sealed lead–calcium-type commonly known as maintenance free. Any of these types can be supplied for starting engines or dc backup in UPS systems. The type used can depend on a number of factors including reliability, life expectancy, cost, and location.

8.6.4 Typical Maintenance Schedules. To assure that emergency and standby power systems perform their vital function, maintenance schedules should be followed. The following schedule is a guide for batteries and chargers in typical service and can vary as local conditions dictate.

Daily:

(1) Check bus voltage

(2) Check pilot cell hydrometer

(3) Check charge or discharge current

Monthly:

(1) Check cell level (add water if necessary)

(2) Check filler cap vent holes for plugs

(3) Inspect electrode plates for color, particles, damage, or swelling

(4) Inspect bottom of cell for residue

(5) Inspect exterior of charger cabinet for signs of water entry

Quarterly:

(1) Perform equalizing charge; after completion, take hydrometer readings of all cells

(2) Read and record cell voltages 10-20 min after starting equalizing charge

(3) Clean off cell tops with soda solution

(4) Check condition of cell terminals and straps; check torque on connection bolts carefully

8.7 Automatic Transfer Switches. Automatic transfer switches require maintenance, as do most components of electrical installations. The automatic transfer
switch is usually applied where two sources of power are made available for the
purpose of maintaining power to a critical load. The necessity of providing safe
maintenance and repair of an automatic transfer switch requires shutdown of both
power sources or installation of a bypass switch. The bypass switch is used to
isolate the automatic transfer switch while maintaining power to the critical load.
For further discussion on bypass switches, see Chapter 5.

8.8 Conclusions. Maintenance is an essential requirement of any electrical installation. The critical nature of emergency and standby power systems applications
dictates the importance of properly maintaining the equipment involved. A good
preventive maintenance program will aid in maintaining a reliable system.

8.9 References. The following publications shall be used in conjunction with this
chapter.

[1] ANSI/NFPA 99-1984, Health Care Facilities Code.

[2] ANSI/NFPA 110-1985, Emergency and Standby Power Systems.

[3] McWILLIAMS, D. W. Maintenance of Emergency and Standby Power Systems.
*Conference Record of the 1976 IEEE Industrial and Commercial Power Systems
Technical Conference*, IEEE 76CH108191A, pp 16–20.

Chapter 9
Specific Industry Applications

This chapter summarizes needs for emergency or standby electric power observed in each of several industries. Industries that have not demonstrated general need for such power are not listed, although there is probably no industry that has not used standby power of some type at some place or time. Even certain general uses of specific applications are ignored. Battery control sources for electrical switchgear are not referenced since they are covered in ANSI/IEEE C37.20-1969 [3].[21] Use of the common flashlight is not referenced unless the need for standby lighting is generally significant.

It is the intent of this chapter to catalog where standby power has been used, to point out how such power can serve to fulfill certain needs, and to provide some guidance in evaluating the desirability of standby power. It is not the intent of this chapter to suggest or interpret legal requirements for standby or emergency power. Such requirements, as may be included in building or safety codes, are adopted and interpreted by authorities having jurisdiction over such matters. The significant reasons for adopting these requirements are far broader than the engineering considerations of the standby power system itself.

Needs for standby power appear to be growing as equipment and process complexities and costs increase. As more industries find that standby power generally can be economically justified, they will be added to the summaries.

It is sometimes difficult to establish the justification for standby power on an objective economic basis. Developing objective historical costing of power outages can be difficult, and projecting costs can be more difficult. Data on power system outages are available in ANSI/IEEE Std 493-1980 [4], but local conditions of weather or utility service can result in performance at considerable variance with the standard data. Infrequency of outages in some locations tends to restrict the data base and the accuracy of short-time predictions. Such data should be augmented with the user's and other local data. The American Insurance Association has adopted the National Building Code [11], which contains requirements for standby based upon types of buildings and occupancy classifications. Compliance with this code is encouraged through adjustment of insurance rates.

Needs for standby power affecting human safety should not be evaluated purely on an economic basis. Standby power should be made available if there is a reason-

---

[21] The numbers in brackets correspond to those of the references listed at the end of this chapter; when preceded by B, they correspond to the bibliography at the end of this chapter.

able possibility that loss of normal power could cause serious injury or loss of life. Emergency or standby power has been mandated by authorities in many situations, and such mandates will likely spread to other situations upon occurrence of injury or death.

Electrical standby power is not recommended as the correct or only solution to all problems or hazards that result from loss of electrical power. There is risk that local electrical failure could forestall use of the standby power. Alternate solutions may be attractive from both an economic and reliability standpoint. For example, cement plants often supply rotational power for the kiln from a mechanically coupled engine during power outages. While avoiding the cost of a generator, this arrangement covers not only loss of utility power, but also the contingencies of loss of in-plant distribution or failure of the electric kiln drive. Many jurisdictions also demand this type of drive as a standby for the electric drive on a chair lift.

Systems that are inherently safe will provide greater reliability than those dependent upon a source of electrical power to maintain safety. Designers are encouraged to produce systems that will assume a safe posture upon loss of electric power. Managers are encouraged to invoke procedures that will avoid hazards when power is lost. In the case of a work force that is required to carry flashlights, or a process controller with built-in battery backup, it may be argued whether they are inherently safe systems or whether they are examples of standby power systems — but they are effective and economically attractive.

Standby power can more easily be justified if generators are also operated in the *cogeneration* mode. Cogeneration refers to obtaining a second form of energy from the generation process, usually heat, which increases the efficiency in the use of energy as well as the economics of the operation. By maintaining certain efficiencies and utilizations, a generator operator may become a qualifying facility under the auspices of PURPA (Public Utilities Regulatory Act of 1978), which will allow them to operate in parallel with the utility. Parallel operation normally increases the economic benefit to be derived from the generators, while allowing *no-break* standby power. Qualifying facility status can make the generator operator eligible for certain tax benefits. With proper design and certain relaying and control additions, the basic cogeneration system can provide standby power, and this is commonly done.

Tables 21 through 30 itemize emergency and standby power needs that are at least somewhat distinctive to the designated industry. There may be needs that are common to many industries, but these are listed only once, under the industry with which they are most commonly associated. For instance, most industries have offices or occupancies to which the items in Table 23 (commercial buildings) would apply. Most industries have some form of data processing, usually with some power supply needs. For larger data processing installations, refer to Table 25 (financial data processing).

Process controllers are proliferating throughout industry, but the tables are intended to address the electrical needs of the process itself rather than the controller. The consideration of supplying power to a controller to prevent malfunction, damage, or loss of data is common to all industries.

Note A
Categories of Purposes for Emergency or Standby Power (for Tables 21-30)

1. Human Safety and Health — This includes preventing injury or death from causes such as falling, collisions, equipment operations or malfunctions, explosions, fire, release of toxic or explosive substances, failure of safety or life supporting devices, and actions of others. Emergency power uses include signaling, lighting, conveyances for evacuation, control of machinery, combating fire, and smoke or vapor dispersal.

2. Environmental Protection — This includes prevention of damage to environment (temporary or permanent), which includes effects on water quality, wildlife, or vegetation that do not immediately or significantly threaten human life or health.

3. Nonemergency Evacuation — This covers requirements of evacuation for convenience or for precautionary reasons where no immediate danger or urgency exists, and can include lighting and conveyances.

4. Protection of Equipment, Product, Material, and Property — This covers needs for orderly shutdown, limited operation, or security, lack of which could result in economic loss due to repair or replacement costs or unavailability of equipment or material.

5. Continuance of Partial or Full Production — This covers needs where cessation of production could result in inordinate economic loss due to extended restart times or requirements of timely product delivery.

Note B
Categories of Types of Standby or Emergency Power Required (for Tables 21-30)

1. Uninterruptible AC Power — Specially conditioned.

2. Uninterruptible AC Power — Suitable for computer use (see 3.10.3 for guidelines to computer power requirements).

3. Interruptible AC Power — Specially conditioned to eliminate sags, spikes, flicker, harmonics, or noise such as could be achieved with static or rotating power conditioner. Required for powering equipment with no inherent immunity to these power system defects.

4. Interruptible AC Power — Commercial grade such as defined by ANSI C84.1-1982 [2] or ANSI/NEMA MG1-1978 [6] motor requirements. Suitable for equipment satisfactorily powered from conventional utility source. Range of tolerable outage time to dictate whether fixed or portable, automatic or nonautomatic.

5. Permanently Installed DC Power Maintained at Full Capability — Includes emergency lighting sources and internal or external battery backup for communications, computers, and processors.

6. Portable DC Power — Includes power sources in flashlights, lanterns, and portable communicating devices, either installed or spare.

Table 21
Summary of Possible or Typical Emergency or Standby Power Needs for the Agri-Business Industry

Power Use	Application	Purposes (see Note A on page 251)	Range of Tolerable Outage Time	Duration of Need	Power Sources (see Note B on page 251)	Remarks
Ventilation, feeding	Confinement, housing	4, 5	1 h or less	Duration of outage	4	
Ventilation, heating, feeding, egg gathering, lighting	Poultry production	4, 5	10 min	Duration of outage	4	Production affected after 10 min. Death can occur after ½ h.
Computer	Data or process control	4 or 5	0–1 h	Duration of outage	3, 4, 2	
Milking machine	Milking	4, 5	1–2 h	2–4 h twice daily	4	Cows must be milked even though milk not saved.
Refrigeration	Food storage	4, 5	½–15 h	Duration of outage	4	Temperature can be maintained for many hours if doors not opened.
Pumps, heater	Water supply systems	4, 5	2–6 h	Duration of outage	4	Alternate solutions available not requiring electric power.

Table 22

Summary of Possible or Typical Emergency or Standby Power Needs for the Cement Industry

Power Use	Application	Purposes (see Note A on page 251)	Range of Tolerable Outage Time	Duration of Need	Power Sources (see Note B on page 251)	Remarks
Kiln drive	Rotation	4	5 min	Indefinite or intermittent, or both	4	Often accomplished by mechanical power.
Control	Shutdown and continuity	4	0	Short or indefinite	2, 5	
Plant	Production	5	0–10 min	Duration of outage	4	Cogeneration using waste heat may be economically justified, particularly with poor commercial supply.

Table 23
Summary of Possible or Typical Emergency or Standby Power Needs for the Commercial Building Industry

Power Use	Application	Purposes (see Note A on page 251)	Range of Tolerable Outage Time	Duration of Need	Power Sources (see Note B on page 251)	Remarks
Emergency lighting	Hotels, apartments, stores, theaters, offices, assembly halls	1, 3, 4, 5	10–60 s or statutory requirement	1½ h or statutory requirement or as needed	2, 4, 5[†]	Building management or safety authorities to determine as needed requirements.
Elevators	Hotels, apartments, stores, offices	3, 5	10 s to 5 min	Evacuation completion	4[‡]	Other methods of evacuation may be available.
Phones in elevators	Signaling, switching, talking	3	10–60 s	Evacuation completion or extended occupancy	2, 4, 5	Many PBX or PAX systems no longer have batteries.
Other phones or PAs	Signaling, switching, talking	3	10–60 s	Evacuation completion or extended occupancy	2, 4, 5	Many PBX or PAX systems no longer have batteries.
HVAC	Space, process, or material	4, 5	1–30 min	Duration of outage	4	Negotiable with tenants.
Tenants' computers	Various	4, 5	Negotiable	Negotiable	2, 3, 4	Negotiable with tenants. For large systems see Table 25.
Fire fighting and smoke control	Fire pumps and ventilation control	1, 4	10 s	2 h or more	4	Pumps needed when high-rise exceeds water pressure capability.
Electric security locks	Access controls	4	0–10 s	Until manual locking accomplished	1, 2, 3, 4	Loss of power should not prevent egress.
Emergency lighting	Store lighting	4	10–60 s	Completion of evacuation	2, 4,[‡] 5[†]	Protection from theft of stock and evacuation.
Emergency lighting and HVAC	Casino lighting and environment	3, 4, 5	1–10 s	Duration of outage	2, 4	Protection from theft and continuing operation are financially significant.

[†]Where applicable, NEC [7] (National Electrical Code, ANSI/NFPA 70-1987), Article 700-12, requires 1½ h battery supply.
[‡]Where applicable, NEC [7], Article 700-12, requires 2 h generator fuel supply.

NOTES: Statements regarding commercial buildings have to be somewhat general due to wide variations in building construction and use, but certain standardization does apply.

Definitions of performance requirements of emergency and standby power systems appear in NEC [7], Articles 700, 701, and 702; and ANSI/NFPA 99-1984 [8]. Standard requirements for use appear in NEC [7], Article 517; ANSI/NFPA 101-1985 [9]; the Uniform Building Code [13]; and the Basic National Building Code [10]. Legal requirements are established by authorities having jurisdiction, often by adopting one or more of the above codes. In addition, conformance with the National Building Code [12] may be made a requirement for insurance by the underwriter. Nevertheless, emergency or standby power is not universally required for all commercial buildings. Notes to NEC [7], Articles 700 and 701, suggest some circumstances that may require such power.

The statutory requirements for standby or emergency power are the minimum permitted. It should not be presumed that they will be adequate for all occasions. If occupants are not evacuated during the mandated period of emergency lighting or power, there may be inadequate lighting for subsequent evacuation, inadequate heating or ventilation, and inadequate water pressure for fire fighting. In addition, tenants may have processes or process equipment, controlled environments, or other needs, some of which are listed in the summary. Thus a loss of power, or the necessity of evacuation if commerical power is lost, could be a hardship. Evacuation of residential facilities during inclement weather could be undesirable. Unless notified or evacuated, tenants may be unaware that commercial power is lost, and therefore unaware of impending loss of power or lighting for evacuation.

Tenants are advised to make themselves aware of power needs during loss of commercial power, and to make themselves aware of the power facilities installed in their building. Tenants with greater needs should either contract with the building management for additional power or supply the necessary facilities themselves. Building managements may find it commercially attractive to offer enhanced or extended emergency power services.

If the requirements are only for emergency lighting, for the minimum 1.5 h specified by NEC [7] battery lights will usually suffice. If lighting needs are more extensive, or if operation of motors is required, an engine generator is normally required. Once this latter equipment is installed, the only limit on the duration of availability of the power is the size of the fuel tank.

Table 24
Summary of Possible or Typical Emergency or Standby Power Needs for the Communications Industry

Power Use	Application	Purposes (see Note A on page 251)	Range of Tolerable Outage Time	Duration of Need	Power Sources (see Note B on page 251)	Remarks
Central office, switching or repeater stations	Traffic transmission	1, 2, 4, 5	0	Duration of outage	3, 4, 5	Battery is primary source with generator backup.
Satellite earth stations	Traffic transmission	1, 5	0	Duration of outage	1, 2, 4, 5	Power requirements largely ac.
Public safety, radio links and terminals	Police and fire communication	1, 2, 3, 4	0–10 s	Duration of outage	2, 4, 5	Usually engine generator.
Military links and terminals	Defense and diplomatic	1, 4	0	Duration of outage	2, 4, 5	Normal requirements, 99.6% reliability.
Aircraft warning lighting	Towers	1	0	Duration of outage	4, 5	

NOTES: The communications industry includes both common and private carriers, and both point-to-point links and broadcast facilities. Traffic can include radio/television entertainment, data transfer, domestic and business voice transmission, and public safety and national defense functions. These facets of communications are grouped, since in many cases all of the previously listed traffic may be transmitted through the same common carrier, whose performance standards must meet the standards of the most exacting traffic.

The immediate purpose of standby power from the carrier's standpoint is continued timely delivery of the product. The purposes of the timely delivery can include protection of life, property, or environment by police, fire, or defense departments. The purpose can include any aspect of business operation, such as marketing, production, purchasing, or financial transactions.

Table 25
Summary of Possible or Typical Emergency or Standby Power Needs for the Financial Data Processing Industry

Power Use	Application	Purposes (see Note A on page 251)	Range of Tolerable Outage Time	Duration of Need	Power Sources (see Note B on page 251)	Remarks
Data processing equipment	Any critical application	4, 5	0	Up to duration of outage	2	UPS can be supported with standby power.
UPS system	Data processing equipment power supply	4, 5	UPS ride-through (usually 10–30 min)	Up to duration of outage	4	This source should meet 10 s requirement if also used for emergency functions.
HVAC	Data processing equipment cooling	4, 5	1 min	Per above	4	To include direct equipment cooling.
UPS system bypass	Power to data processing equipment if UPS fails	4, 5	0	Per above	3, 4	See Chapter 7 and NEC [7], Article 250, for proper grounding. Bypass connection may cause UPS to no longer be defined as a separately derived source.

NOTES: This is an industry where the economic benefits of continued production are substantial enough to justify uninterruptible power supplies, and usually alternate power sources that can maintain operation past the stored energy capability of the uninterruptible power supply (UPS). The capital cost of the large data processing equipment is justified only with optimum usage, and the utter dependence of the industry on this equipment for timely conduct of business would make nonavailability a prohibitive expense.

The first premise in standby power consideration is the need when utility power is lost. It is becoming more apparent that electrical problems on the premises can also interrupt power to the data processing equipment. It is also becoming more apparent that these on-premises problems will proliferate unless equipment can be maintained and repaired.

The above realizations have illustrated the necessity of redundant circuits and power supplies so that power can be maintained during maintenance work or in case of failures. In the latter situation, the system should be designed, and protective devices applied, so as to provide a realistic ability to clear faults selectively.

This industry category includes not only those obvious members, such as banks and thrift institutions, but also any industry whose sales, billing, or other financial operations have caused the assembling of comparable data processing systems.

Table 26

Summary of Possible or Typical Emergency or Standby Power Needs for the Health Industry

Power Use Application	Purposes (see Note A on page 251)	Range of Tolerable Outage Time	Duration of Need	Power Sources (see Note B on page 251)	Remarks
Life safety equipment		10 s		4	Legally required emergency power.
Critical equipment		10 s		4	Legally required emergency power.
Essential equipment		1 min		4	Legally required standby power.

NOTES: These facilities have some of the greatest and the most regulated needs for emergency power for the preservation of human life. This chapter recognizes this industry as a significant user of emergency power, but refers the reader to ANSI/IEEE Std 602-1986 [5], where complete coverage of this topic is provided, and also to Articles 517 and 700 in the NEC [7], plus any local codes or ordinances, for detailed requirements.

Table 27

Summary of Possible or Typical Emergency or Standby Power Needs for the Mining Industry

Power Use	Application	Purposes (see Note A on page 251)	Range of Tolerable Outage Time	Duration of Need	Power Sources (see Note B on page 251)	Remarks
Hoists	personnel	3	1–10 min	± 1 h	4	Requirements vary greatly between sites; safety-related requirements are covered under the CFR, Title 30 [11].
Fans	Ventilation	1	1 min	1 h or more	4	Requirements vary greatly between sites; safety-related requirements are covered under the CFR, Title 30 [11].
Pumps	Drainage	1, 4	1–60 min	Indefinite	4	Requirements vary greatly between sites; safety-related requirements are covered under the CFR, Title 30 [11].

Table 28
Summary of Possible or Typical Emergency or Standby Power Needs for the Petrochemical Industry

Power Use	Application	Purposes (see Note A on page 251)	Range of Tolerable Outage Time	Duration of Need	Power Sources (see Note B on page 251)	Remarks
Purge air supply		1 or 4	10 s	As determined	4, 5	To prevent incursion of combustible or toxic gases into machinery, housed field instrumentation (analyzers, etc), or personnel areas.
Lighting	Emergency evacuation	1	0–10 s	5 min minimum	1–6	Inspection of facilities following a blackout, prior to black start, or evacuation from imminently hazardous areas.
Process controllers	Continued operation or orderly shutdown or both	1, 2, 4, or 5	0–30 min	As determined (by process)	1–5	
Motor operated valves for emergency use	Isolation of fires by blocking flow of hydrocarbons	1, 2, or 4	0	30 s maximum	2, 4, or 5	
Steam generation	Orderly shutdown or continued operation	1, 2, 4, or 5	0 to as determined	As determined	3, 4, or 5	See 3.5 in Chapter 3.
Flame detectors	Orderly shutdown	1, 2, 4, or 5	0	As required for shutdown	1, 2, or 4	
Selected critical process loads	Orderly shutdown or continued operation	1, 2, or 4	0 to as determined	As required by specific process	4	

Table 29
Summary of Possible or Typical Emergency or Standby Power Needs for the Ski Resort Industry

Power Use	Application	Purposes (see Note A on page 251)	Range of Tolerable Outage Time	Duration of Need	Power Sources (see Note B on page 251)	Remarks
Lifts	Uphill transport	3, 5	1–10 min	Until lift is empty — 5–20 min — or for continuous operation	4	Evacuation also can be accomplished using mechanical coupled standby engines. These are required in many jurisdictions.
Slope lights	Night skiing	1, 3	0–1 s	10 s to cease normal skiing; 15–30 min for evacuation	2, 4, 5	Skier can suffer injury from collision or fall with loss of light. Evacuation can tolerate lower light levels.
Lift emergency lighting	Night lift operation	1, 3	0–1 min	Until lift is empty — 5–20 min	2, 4, 5	Required by ANSI B77.1-1982 [1] for unloading or evacuation.

Table 30
Summary of Possible or Typical Emergency or Standby Power Needs for the Waste Water Industry

Power Use	Application	Purposes (see Note A on page 251)	Range of Tolerable Outage Time	Duration of Need	Power Sources (see Note B on page 251)	Remarks
Pumps	Influence, effluence, or processing of waste water	2, 4, 5	0–10 min	Duration of outage	4	EPA or local statutes, or both, define bypassing restrictions.
Process controllers		2, 4, 5	0–10 min	Duration of outage	1, 2, 3, 4, 5	To meet above requirements.
Telephones	Signaling and talking	2, 4, 5	0–1 min	Duration of outage	2, 4, 5	The geographical dispersion in the typical facility requires a facility communication system.

NOTES: (1) Local power generation sometimes accomplished with digestor gas.
(2) Facilities are closely regulated and individually permitted under federal clean water laws. Loss of power could require gravity bypassing of processes that require pumping. These processes may exceed permitted allowances, and in no case is complete bypass allowed. Even permitted abnormal operation may receive extensive unfavorable publicity and may be unacceptable to local citizens. Lack of adequate gravity bypass capability or control thereof could cause flooding. It is usually necessary to maintain level, flow, and temperature monitors to oversee bypassing or other contingency operations.

Minimum standby power should be whatever is necessary to meet permit requirements. Many plants have achieved 100% standby capability through the use of digestor gas to operate generators.

Communication should be maintained to coordinate efforts involving manual control or monitoring at various locations. Many modern PBX systems depend upon commercial power for operation.

References. This standard shall be used in conjunction with the following publications:

[1] ANSI B77.1-1982, American National Standard Safety Requirements for Passenger Tramways — Aerial Tramways and Lifts, Surface Lifts, and Tows.

[2] ANSI C84.1-1982, American National Standard Voltage Ratings for Electric Power Systems and Equipment (60 Hz).

[3] ANSI/IEEE C37.20-1969 (R 1982), IEEE Standard for Switchgear Assemblies Including Metal-Enclosed Bus.

[4] ANSI/IEEE Std 493-1980, IEEE Recommended Practice for Design of Reliable Industrial and Commerical Power Systems.

[5] ANSI/IEEE Std 602-1986, IEEE Recommended Practice for Electric Systems in Health Care Facilities.

[6] ANSI/NEMA MG1-1978, Motors and Generators.

[7] ANSI/NFPA 70-1987, National Electrical Code.

[8] ANSI/NFPA 99-1984, Health Care Facilities Code.

[9] ANSI/NFPA 101-1985, Life Safety Code.

[10] Basic National Building Code, Building Officials and Code Administrators (BOCA International, Inc).[22]

[11] CFR (Code of Federal Regulations), Title 30: Mineral Resources (parts 1–199), Mine Safety and Health Administration, Department of Labor, July 1986.[23]

[12] National Building Code, American Insurance Association.[24]

[13] Uniform Building Code, International Conference of Building Officials.

---

[22] This publication is available from BOCA International, Inc, 4051 West Flossmoor Road, Country Club Hills, IL 60477-5795.

[23] This document is available from the Superintendent of Documents, US Government Printing Office, Washington, DC 20402.

[24] This publication is available from the Publications Department, American Insurance Association, 85 John Street, New York, NY 10038.

Index

Acknowledgments

Appreciation is expressed to the following groups for their valuable contributions during the preparation of this document:

Electrical Generating Systems Association (EGSA)

Institute of Electrical and Electronics Engineers (IEEE): Power Engineering Society, Power Generation Committee, and Communications Society, Transmission Systems Committee

Appreciation is expressed to the following companies and organizations for contributing the time of their employees to make possible the development of this document:

Automatic Switch Company
Chrysler Corporation
Computer Power Products
Constant Power Systems
Cummins Engine Company, Inc
El Paso Natural Gas Company
E. I. Dupont De Nemours & Company
Exxon Research and Engineering Company
Flour Technology, Inc
FMC Corporation
Kohler Company
LCR Consulting Engineers, P. A.
Malcolm Pirnie, Inc
McDonald Equipment Company
Oak Ridge National Laboratory
Onan Corporation
Sandia National Laboratories
SOHIO Alaska Petroleum Company
Solar Turbines Incorporated
The University of Alberta
US Department of Agriculture
Westinghouse Electric Corporation